中英对照

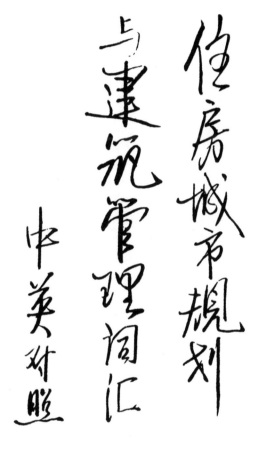

住房城市规划与建筑管理词汇

中英对照

住房城市规划与建筑管理词汇项目组　编

商务印书馆
The Commercial Press

2016年·北京

图书在版编目(CIP)数据

住房城市规划与建筑管理词汇/住房城市规划与建筑管理词汇项目组编.—北京:商务印书馆,2016
ISBN 978-7-100-10530-9

Ⅰ.①住… Ⅱ.①住… Ⅲ.①住宅建设—城市规划—词汇 ②城市规划—施工管理—词汇 Ⅳ.①TU984.12-61

中国版本图书馆 CIP 数据核字(2013)第 318419 号

所有权利保留。
未经许可,不得以任何方式使用。

住房城市规划与建筑管理词汇(中英对照)
住房城市规划与建筑管理词汇项目组 编

商 务 印 书 馆 出 版
(北京王府井大街36号 邮政编码100710)
商 务 印 书 馆 发 行
北京新华印刷有限公司印刷
ISBN 978-7-100-10530-9

2016年10月第1版　　　开本 787×1092　1/16
2016年10月北京第1次印刷　印张 50¼
定价:128.00元

出版说明

中国实行改革开放五年后的1983年,中美两国的政府机构——中国城乡建设环境保护部(即现在的住房和城乡建设部)和美国住房与城市发展部,为了减少双方交流时因对住房、城市规划和施工管理等词义的理解不同而引起的误解,决定共同编写一本词汇。1987年,这项合作成果以《住房城市规划与建筑管理词汇》(以下简称《词汇》)的形式结集成册,并在行业内推广使用,成为正确解读、理解中美两国有关文献的重要依据。由于种种原因,这份凝聚了中美双方多位专家心血的《词汇》,未能在当年及时正式出版发行。随着中国住房市场化改革和建筑业科研院所的改制转型重组,加上原项目组中方专家的陆续退休,《词汇》的出版工作被长期搁置。

2013年,经中国建筑设计研究院2007级研究生赵希、2011级研究生万子昂完成录入整理,北京建筑大学2014级研究生张笑轩重新进行汉字注音,由中美合作项目组中方专家张珑女士审校全文并撰写后记，中国可持续发展研究会人居环境专业委员会何建清、国家住宅科技产业技术创新战略联盟潘晓棠共同策划，中国建筑设计研究院盛况联系商务印书馆，《词汇》终于在项目启动30周年之际得以付梓。

  作为中美政府合作项目的参加单位，改制重组后的中国建筑设计研究院（集团）长期以来将《词汇》作为本领域国际合作交流、工程咨询投标用词的重要技术依据，并作为本院"建筑设计及其理论"方向研究生培养的辅助教材。《词汇》的出版发行，既标志着中美城市和建筑领域合作基础的扎实，也标志着中美两国在城市精明增长、健康社区建设、绿色建筑推进等方面进一步开展深入交流与合作的发展空间的广阔。同时，中国正处在快速城镇化进程中，以保障性住房建设为代表的第二次住房改革正在按计划进行，《词汇》的出版发行，一方面有助于我国学习和借鉴国际发展经验，另一方面也有助于将我国的实践经验及时、准确地传达给国际社会。《词汇》的出版发行，必将在传播国际城市和建筑文化、消除政治误解和增进社会交流中起到积极的促进作用。

何建清

2013 年 7 月 28 日于北京

ENGLISH – CHINESE GLOSSARY OF TERMS

IN

HOUSING, URBAN PLANNING,

AND CONSTRUCTION MANAGEMENT

United States of America
Department of Housing and
Urban Development (HUD)

People's Republic of China
Ministry of Urban and Rural
Construction and Environmental
Protection (MURCEP)

Washington, D.C. - Beijing
1987

住房城市规划
与建筑管理词汇
中英对照

中华人民共和国
城乡建设环境保护部

美利坚合众国
住房与城市发展部

中国北京·美国华盛顿
一九八七年

This Joint Venture

is respectfully dedicated to

谨将此合作项目

报 给

The Secretary of Housing and Urban
Development of the United States
of America

Samuel R. Pierce, Jr.

The Minister of Urban and Rural
Construction and Environmental
Protection of the People's Republic of China

Ye Rutang

美利坚合众国
住房与城市发展部部长

塞缪尔·皮尔斯

中华人民共和国
城乡建设环境保护部部长

叶如棠

目 录

前言　　/ xiii

编写说明　　/ xv

致意　　/ xxix

词汇正文　　/ 1

主题分类　　/ 682

参考书目　　/ 708

汉语拼音对照表　　/ 710

编写《词汇》的点滴回忆　　/ 744

FOREWORD

The English-Chinese glossary of housing, urban planning, and construction management terms is designed to help overcome language barriers, to enhance mutual understanding, and to facilitate future exchanges between the People's Republic of China and the United States. The project was undertaken under the Protocol on Cooperation between the Department of Housing and Urban Development (HUD) of the United States and the Ministry of Urban and Rural Construction and Environmental Protection (MURCEP) of China. This is but one of several protocols in different fields under the U.S.-China Agreement on Science and Technology. On the occasion of the publication of the glossary, we would like to express our sincere hope that this work will serve as the cornerstone for future efforts and that cooperation between the United States and China will prove even more fruitful in the years to come.

U.S.- China Program Co-Chairmen

Ambassador Theodore R. Britton, Jr.
Assistant to the Secretary for International Affairs
Department of Housing and Urban Development (HUD)

Xu Ronglie
Engineer-in-Chief
Ministry of Urban and Rural Construction and Environmental Protection (MURCEP)

前 言

中英对照《住房城市规划与建筑管理词汇》作为中美科技合作协定的项目之一,是根据中华人民共和国城乡建设环境保护部和美利坚合众国住房与城市发展部的科技合作协定书编写的,旨在克服语言障碍,沟通思想,增进了解,促进今后的交流合作。值此出版之际,我们衷心祝愿中美两国科技合作结出更为丰硕的果实。

中美联席主席

许溶烈

中国城乡建设环境保护部
总工程师
许溶烈

美国住房与城市发展部
外事部部长助理
西奥多·布立顿大使

EDITORS' INTRODUCTION

This English-Chinese glossary is not just a list of words and definitions in the two languages. Rather, it strives to provide the discriminating reader with insights into the latest ideas and developments in specialized fields in China and the United States.

The glossary should not be viewed as a final product, but as a basic first step in improving communications between the two countries in the planning and urban development fields. The definitions are based on information from many dictionaries and glossaries, which has been modified, when necessary, according to current urban development practice. Every attempt was made to use authoritative definitions, if they existed.

In both countries, words are rooted in a rapidly changing professional and national experience. Professional dictionaries and glossaries are soon outdated by changing practice. For that reason, comments on the currency and accuracy of the glossary are welcomed by both countries. We sincerely hope that the years of combined efforts to compile this glossary will reduce the errors and misunderstandings which come from a literal translation of Chinese and American terms. This bilingual reference work should pave the way for future Chinese-

编写说明

这本中英对照的《词汇》并不仅仅是简单地用两种文字对一些用词作出解释，而且也是力求为细心的读者提供一个认识中美两国所从事的新工作与思想交流的窗口。

《词汇》的出版不是双方合作的终结。恰恰相反，它是两国为在城市规划与建设方面增进了解而迈开的新的一步。词条的解释是根据现有词典的定义，并参照实际情况和经验而作出的。部分词汇的解释则尽量选用有权威性的说法。

这些由两国专家所选入的词，其含义会随着行业及其在两国的不同发展而变化。这种迅速的变化往往使某些专业性的词典显得过时。因此，我们欢迎读者对这本《词汇》中各词条的解释提出宝贵意见，衷心希望经过双方多年努力编出的《词汇》有助于减少因生硬翻译而造成的误解，并为今后中美双方的合作起到促进作用。同时也希望这一成果将有助于增进其他国家对中美两国在城市建设及建筑施工的理论与实践方面的了解。

American cooperation and should also be helpful to other countries in understanding Chinese and American urban development theory and practice.

The glossary was created as one of the projects in an exchange program between the U.S. Department of Housing and Urban Development (HUD) and the Chinese Ministry for Urban and Rural Construction and Environmental Protection (MURCEP). The framework for the project is the bilateral Protocol on Cooperation in the Field of Building Construction and Urban Planning. Work on the glossary project began in 1983, as working groups in Beijing and Washington compiled separate lists of terms and began the complex process of culling down over 1700 terms and composing accurate definitions in Chinese and English.

One of the difficult tasks faced by both working groups was determining the number of terms to be included in the glossary. Suggestions ranged from 300 to 3000 words. The decision was made by both sides to keep the number of words at a manageable level – around a thousand terms – and to concentrate on ambiguous "soft" terms not covered adequately in the many technical dictionaries in the field. Highly technical construction terms were avoided because they are readily found in architectural and engineering glossaries.

The purpose of this glossary is to provide a communi-

本《词汇》是根据中华人民共和国城乡建设环境保护部和美利坚合众国住房与城市发展部所签订的房屋建筑与城市规划科技合作议定书进行的一个合作项目。编写工作于1983年正式开始，先由双方编写组分别拟出词汇共1700条，然后反复筛选，并分别写出中文和英文的解释稿。

编写过程中的一大难题是确定收入词条的数量。双方建议的数量从少至300条到多达3000条。最后协商同意，将数量控制在比较切实可行的1000条左右，主要选入难译词或在一般技术词典中难以解释清楚的"软词汇"。对在建筑施工与工程专业词典中能查到的技术词汇，一般不予考虑。

编写本《词汇》的目的，是为使用汉语和英语

cation tool useful to both English-speaking and Chinese-speaking experts in housing, urban planning, construction management, and related fields. Every attempt has therefore been made to weigh the relative importance of terms to both sides and to retain terms with definitions specific to the field rather than generic. In addition, the terms that have been selected are accepted and commonly used internationally, or are culture-specific to the United States or China. The country of origin of the term is specified in the definition: if no country of origin is noted, the term as defined is commonly used internationally. Most of the 1036 terms selected are practical in nature and the range of the glossary is broad both in subject matter and language level, encompassing theoretical concepts such as "Central Place Theory," construction, jargon such as "stickbuilt," and concepts pertinent to calculation of construction costs in China, e.g., "norm." Slang and vernacular terms, as well as specially coined translations of foreign concepts, are enclosed in quotation marks.

Production of a specialized bilingual glossary by two groups working on opposite sides of the world entailed a variety of formidable challenges. On the broadest level, the working groups were faced with the complex problem of trying to define differences in the meanings of terms that reflect rapid changes in two dynamic cultures, and to match terms in one language to adequately comparable ones in the other. Frequently, misunderstandings and mismatches became apparent only during lively face-to-

的专家和科技人员提供一个在住房、城市规划和施工管理的有关领域里沟通思想、克服语言障碍的工具。为此，双方编写组对打算收录的每一个词进行反复斟酌，只保留符合上述选入条件的词，含义一般的则予以排除。凡中方特有或美方特有的词汇都在解释中予以注明，未经注明者均为国际通用词。最后选定的1036个词条，大都比较实用，涉及面也比较广泛，既有理论概念，如"中心地带理论"；又有专业术语，如"木架房屋"；还有中国用以计算工程成本的"定额"等。凡是俚语、土语以及对方特有的用词的新编翻译都用引号表示。

处于地球两面的两个编写组合作编写一本两种语言对照的专业词汇，是一项难度很高、充满挑战的工作。对于这一大批反映处于不断发展中的两种不同文化的词，既要写出恰当的解释以说明在两国用法的区别，又要做到词面相当，词义相符，确是十分艰难的任务。往往是在面对面讨论以至激烈争辩之后，才发现原来对某些词的理解与解释是错误的。

face discussions.

The glossary terms are listed in English alphabetical order, with the English and Chinese entries placed side by side. A Pinyin index is appended. The glossary working groups have sought to find exact or near equivalents for terms, and in cases where no suitable equivalent existed, to coin appropriate terms in English or Chinese. An example is the Chinese term "borrowed view." The term, "yijing," which is basic to Chinese landscape architecture, has been rendered in its Pinyin form, as the concept is virtually intranslatable into English. When the meaning of a term varies between the two languages, as in the case of "comprehensive planning," separate definitions are supplied. In some instances, a general definition is provided, followed by a description of variant meanings in the two countries. In a few cases, very similar terms have been included , one for each country, because of differences in how they are used and what they connote, e.g., "civic building" and "civil building." The former is a municipal government building and the latter can be any public use building, e.g., a community center or residential building.

Many difficult and humorous problems were encountered in translating American jargon into Chinese. For example, the term "balloon mortgage" was rendered in Chinese as "special big tail mortgage," which conveys the intended meaning of the term. A literal

所收词条均按英文字母顺序编排，英汉对照排版，后附汉语拼音索引。词条尽量采纳准确的译词，或在某种情况下采用词义相当的译法。有些词并无现成译法，如中方词条"借景"，在英语中并无相应的词，只能新编；而中国园林建筑学的基本概念"意境"一词，由于没有相应确切的英文术语，则决定保留其汉语拼音。有些词的含义在两国不同，如"总体规划"，则根据不同的实际情况做出两种不同的解释。有些词条的解释则先列出一般用法，然后说明两国用法之差异所在。也有一些词虽然词面相似，而含义却迥然不同。例如"civic building"和"civil building"，前者为"市政建筑"，后者则为"民用建筑"。这也正是把它们收入《词汇》中的原因所在。

在编写中，我们也遇到一些既困难又有趣的问题。例如，把"balloon mortgage"译为"特大尾数抵押贷款"，可以比较确切地表示出它的含义，但如直译为"气球式抵押贷款"，则将令人不知所云。把"brainstorming"译为"绞脑汁会"，可以基本

translation of the term in Chinese, "mortgage in the form of an air balloon," would bring the reader's mind to a standstill. "Brainstorming" was translated into Chinese as "squeezing-the-brain-juices-session." One literal translation into Chinese resulted in "brain concussion."

Certain commonly used American terms selected for inclusion by the Chinese working group proved especially difficult to translate, for example, "amenity," "neighborhood," and "community," because they have a broad range of meanings. Other American terms, such as "bottom line," "underwriting," and "sale leaseback" were included because they are difficult to understand in the Chinese context. Terms were also offered for the glossary that are special to China, such as "squeeze-in development" and "only-child family." The suggestions offered by the 250 organizations and individuals throughout China asked to comment on this glossary proved very useful.

An additional linguistic problem is the confusing number of English and Chinese language variants in use and in print throughout the world today. To reduce the problems of usage, we have taken American English as our standard. The Chinese in the glossary is that used in the People's Republic of China.

Certain closely related terms have been explained together to give a brief overview of a particular system. In such cases, a "see-reference" instead of a definition

达到词面相当和词义相符,但如按字面译成"脑震荡",可就贻笑大方了。

中方所选的某些美国常用词汇,因词义多变而不易确切把握,如"amenity"(宜人事物)、"neighborhood"(邻里)和"community"(社区)等。还有一些较难理解的美国词汇,如"bottom line"(净收益栏)、"underwriting"(风险担保)、"sale leaseback"(出售租回)等。还有部分中国特有词汇,如"见缝插针"(squeeze-in development)、"独生子女户"(only-child family)等。为了避免望文生义的误解,亦予收入。在编写中,中方曾将初稿发至全国250个单位和个人征求意见,收到了不少有益的建议。

鉴于世界各地英语的用法有一定差异,本《词汇》中除个别情况外,一律采用美国英语。中文则采用中华人民共和国通行的用法。

凡成组相关的词条,仅对主要词条作较详细解释,其他词条则以"见……(词条)"的方式处理。与内容有联系的词条,则采用相互"参见"的方式。

under the secondary term sends the user to a key term. Furthermore, interconnections among closely related terms are indicated through the use of cross-references ("See also…") at the end of the definitions.

To help the reader find particular types of terms, eleven subject indexes are provided for the following areas: Housing Types and Uses, Real Estate Finance and Residential Mortgages, Construction Management – Process and Personnel, Construction Estimates and Costs, Bids and Contracts, Area/ Density Measures, Land Use Planning and Zoning, Economic and Community Development, Urban Development Forms, Urban Design and Historic Preservation, and Environment/Infrastructure/Transportation. Acronyms commonly use in American English for certain glossary terms are listed with a "see-reference" to the fully spelled-out terms (e.g., HVAC–See HEATING, VENTILATING, AIR CONDITIONING).

This bilingual reference work was undertaken with the intention of strengthening Chinese-American cooperation in housing construction and urban planning projects. The editors hope that the publication of this volume will also serve to expand the international exchange of ideas on the improvement of the urban environment.

为了方便读者查找专类词汇，我们还就部分词条做了主题索引，包括"住房类型与用途"、"房地产资金与住房抵押贷款"、"施工管理——程序与人员"、"施工预算与成本"、"投标与合同"、"面积、密度"、"土地利用规划与区划"、"经济与社区建设"、"城市发展形式"、"城市设计与历史性建筑保护"以及"环境、基础设施、交通运输"。对于美国常用的首字母缩略词，则用"见'该词'"的办法处理，如"HVAC"（Heating, Ventilating, Air Conditioning）用"见'供暖、通风、空调'"。

我们编写这本双解《词汇》的目的在于加强中美双方在房屋建筑和城市规划领域的合作。双方编辑希望它的出版能够促进国际间在改良城市环境方面的信息交流。

Co-Editors

John M. Geraghty
Office of International Affairs
Department of Housing and Urban Development (HUD)

Lin Zhiqun
Director, Bureau of Urban Housing
Ministry of Urban and Rural Construction and
Environmental Protection (MURCEP)

双方编辑

林志群

中国城乡建设环境保护部
城市住宅局局长
林志群

美国住房与城市发展部
外事办公室
葛瑞蒂

ACKNOWLEDGEMENTS

This glossary was developed jointly by a China working group headed by Mr. Lin Zhiqun in Beijing and a U.S. working group headed by Mr. John M. Geraghty in Washington. The members of the Chinese working group were Ms. Zhang Long, Mr. Sun Huasheng, and Ms. Sun Hanfu. Mr. Chen Zhanxiang, Mr. Sun Zengfan, and Mr. Le Baiyong were advisors to the group. The members of the U.S. working group were Ms. Kathleen M. Dell'Orto, Mr. Jack Underhill, and Mr. Frank Lee. Mr. Donald Fairman has served as a valued advisor. Assistance was also provided at various stages of the project by Mr. William Heenan, Ms. Susan C. Judd, Mr. Tsou-liang Tang, and Mr. Richard Metcalf.

We would especially like to thank two persons who were essential to the task:

Ms. Kathleen M. Dell'Orto
Translator/Lexicographer
U.S. Patent and Trademark Office, and

Ms. Zhang Long
Senior Editor
China Building Technology Development Center

致 意

本《词汇》由林志群领导的中方编写组和葛瑞蒂领导的美方编写组共同完成。中方编写组包括张珑女士、孙骅声先生、孙汉甫女士,陈占祥先生、孙曾蕃先生、乐百墉先生担任中方编写组顾问。美方编写组包括杜凯琳女士、安德熙先生、李夙炯先生,唐纳德·范门先生为美方编写组顾问。威廉·希能先生、苏珊·裘德女士、唐作梁先生以及理查德·麦特卡夫先生在美方的工作进程中也都提供了帮助。

我们特别要向做出重要贡献的中国建筑技术发展中心副编审张珑女士和美国专利与商标局翻译及词典专家杜凯琳女士表示衷心的谢意。

The Chinese side would like to acknowledge the organizations and individuals that have given full support to the work, including the China Building Technology Development Center, the China Academy of Urban Planning and Design, Mr. Shang Kuo, and Mr. Li Changhong. Acknowledgement is also made of all the building design and research institutions and other related organizations throughout the country for the comments and suggestions they offered during the preparation of the drafts.

The U.S. side gratefully acknowledges the valuable information and organizations, including Mr. John Courtney and Mr. Benjamin Cukok, The World Bank, Washington, D.C.; Mr. Weiming Lu, Executive Director, Lowertown Redevelopment Corporation, St. Paul, Minnesota; Mr. David Syphard, CRS Sirrine, Houston, Texas; Mr. Eric Carlson, Institute for Public Administration, New York, N.Y.; Mr. Warren Lasko, Executive Vice President of the Mortgage Bankers Association of America, Washington, D.C.; the information specialists of the U.S. Bureau of Statistics; the staff of the HUD Library; and the American Institute of Architects.

Ms. Harriett Holly and Ms. Jo Ellen Plummer of HUD and Ms. Hoai Hong of The World Bank typed many drafts of the text. Their long, patient hours on the details are much appreciated. The typing of the final Chinese language text by Eastern Computers, Inc., of Virginia Beach was

在编写过程中，中方工作组得到了下列单位及个人的大力支持：中国建筑技术发展中心、中国城市规划设计研究院、尚廓先生、李昌鸿先生，在此特致以深切谢意。在草稿的形成阶段，中方曾发函至全国科教和设计单位征求意见，并收到大量有益的建议，特此一并致谢。

美方工作组对下列提供宝贵资料和给予大力支持的单位及个人致以衷心的谢意：华盛顿世界银行约翰·考尼先生及本杰明·丘柯先生、明尼苏达州圣保罗下城改建公司执行委员卢伟民先生、得克萨斯州休斯敦市施雅锡集团大卫·西法尔德先生、纽约市公共管理研究院艾立克·卡尔逊先生、美国抵押贷款银行协会常务副理事长华仑·拉斯科先生、美国国家统计局信息资料专家、美国住房与城市发展部图书馆以及美国建筑师学会。

我们也感谢长时间耐心地承担打字任务的美国住房与城市发展部的海丽爱·郝里女士和乔爱仑·波勒曼女士，以及世界银行的洪皓爱女士。中文稿的最后一次打字由弗吉尼亚海滩的东方电脑公司承担，并在弗吉尼亚州密勒与史密斯住宅建造公

made possible through a generous grant from Mr. Gordon Smith, President, Miller & Smith, one of the major home builders in northern Virginia.

Mr. Walter Stewart, a resourceful Chinese linguist, was of invaluable assistance in resolving the myriad problems of language and presentation that arose in the late phases of the project. Ms. Jina Bang, a gifted graphic artist, did the layout including pasting-up the bilingual texts.

The line drawings of buildings in China were supplied by Mr. Shang Kuo. The drawing of American buildings are taken from Clues to American Architecture, copyright c 1985,1986 by Marilyn W. Klein and David P. Fogle, illustrations copyright c 1985,1986 by Wolcott B. Etienne, and are reproduced here with the kind permission of Starrhill, Inc.

We, the editors, have benefitted immensely from the comments and insights of all advisors, but we assume full responsibility for any errors or misunderstandings contained in the glossary.

司总裁戈登·史密斯先生慷慨解囊相助之下才得以完成。

汉语专家沃尔特·斯图尔特先生对解决本项目最后一阶段的语言与排版相关的若干问题提供了大力帮助。绘画艺术家彭金娜女士承担了版面设计和中英文稿的拼版工作。

中方所用的插图是由尚廓先生提供的。美方在征得斯塔希尔出版社同意后，采用玛丽莲·克兰恩和大卫·福格所著《美国建筑介绍》一书中的部分插图，它们是华尔考·艾汀纳先生的作品，在此谨致谢意。

此项《词汇》编写工作之得以完成，是与各方所给予的建议和帮助分不开的。本《词汇》肯定存在缺点，任何批评、指正与建议将受到衷心欢迎。

A

Abandonment

In the United States, relinquishment by a landlord or owner/occupant of all rights and responsibilities of owner-ship, including collection of rents and provision of services to tenants. Abandonment frequently occurs in poor inner city areas with high crime rates.

Accelerated depreciation

In the United States, permission of the federal government entitling a taxpayer to claim greater-than-usual losses in the value of income-producing property for the years immediately after purchase rather than depreciating the property at a steady rate over its useful life. Tax authorities allow this practice to encourage investment. See also DEPRECIATION.

Acceptance of hidden subsurface work

In China, approval of portions of construction work that will be covered up or otherwise inaccessible when the project is completed, e.g., inspection and acceptance is requisite to undertaking further construction. See also FINAL ACCEPTANCE.

Acceptance of work subelements

In China, approval of certain defined subdivision of construction work. Acceptance procedures are usually carried out upon completion of the main structural

放弃产权 Fàngqì chǎnquán

在美国，指房产主或业主对其具有所有权的产业放弃一切权利和义务，包括收租及对租户提供服务。此种现象通常发生在贫困的犯罪率高的城内衰落区。

加速折旧 Jiāsù zhéjiù

美国联邦政府为了鼓励投资，规定纳税人对产生收益的房地产，允许在购入的初期申报较高的折旧率，即高于其使用期内的固定折旧率。[参见"折旧、贬值"]

隐蔽工程验收 Yǐnbìgōngchéng yànshōu

在中国，指对某些竣工后不便检查的工程项目（如地下的基础、基础下的桩、钢筋混凝土中的钢筋等），在该工程施工完毕，下一工序开始之前，及时组织的验收。[参见"最终验收"]

分项工程验收 Fēnxiànggōngchéng yànshōu

在中国，指对已完工的某一分项工程的验收。通常对于一般工程项目的主体结构工程，以及对于重点工程项目、特殊工程项目或采用新技术的工

work in an ordinary project, or upon completion of all work subelements in a key project, special project, or project employing innovative techniques. See also COMPLETED ITEM, CONSTRUCTION WORK ELEMENT, and FINAL ACCEPTANCE.

Accepted bid

In the United States, the bid or proposal approved by the owner or his representative as the basis for a contract for the proposed construction. See also BID, CONTRACT AWARD, and CONSTRUCTION COST.

Access right

In the United States, right of a landowner to pass to and from his/her land, usually by a dedicated street or easement across another landowner's property. See also EASEMENT and RIGHT-OF-WAY.

Accessory building

A building located on a lot and used for a purpose related indirectly to that of the principal building on the same lot, e.g., a garage or storage shed. See also ACCESSORY USE.

Accessory housing

In the United States, a single-family house that has been divided into self-contained units. While the

程项目的全部工程进行分项工程验收。[参见"建成投入生产项目"、"分部工程"、"最终验收"]

中选标 Zhòngxuǎn biāo

在美国,指业主或其代理人为拟建工程所选中的标单,作为准备与投标单位签订承包合同的基础。[参见"投标"、"发包"、"建筑成本"]

出入权 Chūrùquán

在美国,指地产主出入其所拥有土地的权利,通常指经过指定街道或另一地产主的产地。[参见"地役权"、"通行权"]

附属建筑 Fùshǔ jiànzhù

主要建筑以外的辅助建筑,即在同一基地上做其他有关用途的建筑,例如车库、存贮棚等。[参见"附带用途"]

合住住房 Hézhù zhùfáng

在美国,指原供一家居住的住房被划分为几套独立的居住单元,这在许多地区是非法的。这种做

practice is sometimes prohibited, many jurisdictions are amending their regulations because of the need for affordable housing.

Accessory use

In the United States, use of land or a secondary building for purposes incidental to the primary use of the same lot. Accessory uses are frequently defined in zoning ordinances. See also ACCESSORY BUILDING, PERMITTED USE, and ZONING.

Accreditation

Official recognition of the competence of an educational institution or testing laboratory by a public or private organization after evaluation according to accepted criteria or standards. See also BUILDING MATERIAL TESTING and CERTIFICATION OF MATERIALS AND PRODUCTS.

Acquisition approval

In China, approval of land acquisition granted by the urban planning administration to a development organization for construction agreed to by the planning commission. Factors considered in approval decisions include the size, dimension, and location of the site within a city or its suburbs. See also LAND ACQUISITION.

法虽被禁止,但有些地区为了满足对住房的需求正在对有关规定进行修改。

附带用途 Fùdài yòngtú

在美国,指同一基地上作次要目的使用的土地或建筑,在区划法规中一般对此种用途有所规定。[参见"附属建筑"、"许可使用"、"区划"]

确认 Quèrèn

由官方或私营组织根据公认的标准对一个教育机构或实验室的检验能力作出评价后予以认可。[参见"建筑材料检验"、"材料与制品鉴定"]

拨地 Bōdì

在中国,指城市规划管理部门对于业经计划部门批准的建设项目,按总体规划和详细规划的规定拨给建设单位建设用地,包括用地的地址、范围、周边尺寸等。[参见"征地"]

Activity analysis

Studies of time and space patterns followed by households, firms, and various other institutional entities in conducting their affairs. Activity analysis is potentially useful in special planning exercises.

Adaptive re-use

The process of transforming an old structure, usually either abandoned or underutilized, into something quite different from the purpose for which it was originally built, e.g., redesigning an abandoned railroad station as a shopping mall. Examples of adaptive re-use have become prominent features in redevelopment areas of many U.S. cities. See also URBAN REDEVELOPMENT.

Addendum

In the United States, a written or graphic instrument issued prior to the execution of the contract, modifying or interpreting the bidding documents-including drawings and specification-by additions, deletions, clarifications, or corrections. As such, addenda become part of the contract documents when the construction contract is executed. See also BIDDING DOCUMENTS and CONTRACT DOCUMENTS.

活动分析 Huódòng fēnxī

关于家庭、商号及其他多种机构在时间和空间上的活动方式的研究。在空间规划工作中极为有用。

适应性再使用 Shìyìngxìng zàishǐyòng

指改造废弃或使用不当的建筑物的过程。改造后的使用目的与当初建造时迥然不同。例如，将废弃的火车站改建为购物商场。在美国，适应性再使用已成为许多城市中地区改造的特色之一。［参见"城市改造"］

补充文件 Bǔchōng wénjiàn

在美国，指签订承包合同之前，业主所补发的文字或图纸文件，通过增删、澄清或改正对招标文件（包括图纸与说明书）进行修改或解释。在以后订立合同时，这种文件即作为合同文件的一部分。［参见"招标文件"、"合同文件"］

Addition

Any construction that increases the floor area or size of a building, such as an additional bedroom, a porch, or an attached garage of carport. See also BUILDING ALTERATION and HOME IMPROVEMENT.

Adjustable rate mortgage (ARM)

In the United States, a new, complex form of mortgage with a variable interest rate linked to a standard financial index. The rate variation may affect the size of monthly payments, the term of the loan, or some combination of these elements. Mortgage plans may have limitations on the size of the periodic or cumulative interest charges or on the permissible amount of changes in payments. Also called a variable rate mortgage. See also MORTGAGE.

Administrative complex of a factory

In China, a group of buildings on a factory site, containing administrative offices and health, library, recreation, and/or dining facilities for the use of employees, as well as green spaces and parking spaces. The complex is usually located near the factory entrance.

Administrative enclave

In China, a separate, noncontiguous area located beyond the contiguous boundaries of a province,

加建工程 Jiājiàn gōngchéng

指扩大建筑面积或增建附加在建筑上的其他工程，例如加建卧室、门廊或汽车库等。[参见"改建工程"、"住房改善"]

可调利率抵押贷款 Kětiáolìlǜ dǐyā dàikuǎn

美国的一种新型、复杂的抵押方式，其利率可随标准金融指数变动。利率的变动将会影响每个月的还款额或贷款的条款，或综合影响这两方面。抵押条款中可对定期的或累计的利息或对付款的允许变动数量加以限制。又称可变利率抵押贷款。[参见"抵押贷款"]

厂前区 Chǎngqiánqū

中国工厂中常见的一组建筑，包括行政管理和职工福利设施，如办公室、卫生所、图书室、文娱活动场所、食堂、绿地及停车场等，一般都设在工厂的入口处附近。

飞地 Fēidì

在中国，指省、市或县内一块土地属他省、市、县所有。例如江苏省的梅山，完全是由上海市提

municipality, or county, but under its jurisdiction. For example, Meishan in Jiangsu Province has developed into a large steel base with financial and technical resources solely from Shanghai, and Meishan's output is counted as part of the total output of Shanghai.

Advance payment for materials

In China, payment advanced to a construction organization for the normal procurement and storage of necessary materials, to be deducted in installments from regular payments for work progress.

Advanced wastewater treatment (AWT)

In the United States, the final stage of wastewater treatment, involving removal of nutrients such as phosphates and nitrogen, as well as any remaining suspended solids. See also PRIMARY WASTEWATER TREATMENT, SECONDARY WASTEWATER TREATMENT, SEWAGE DISPOSAL SYSTEM, WASTE MANAGEMENT, and WASTE UTILIZATION.

Advocacy planning

In the United States, the preparation of plans or planning proposals and their advocacy by professional planners on behalf of an organization, interest group, or community as an alternative or

供资金和技术力量而建设成为大型的钢铁基地，其产量为上海市生产总量的一部分。

预付材料款 Yùfù cáiliàokuǎn

中国的建设单位预付给施工单位的备料款，用以解决正常的材料储备费用。这项预付款，在工程开工后支付工程进度款时分次扣回。

污水深度处理 Wūshuǐ shēndù chǔlǐ

在美国，指污水处理的最后阶段，包括去除营养物质，如磷酸盐和氮，以及剩余的悬浮物质。[参见"污水一级处理"、"污水二级处理"、"污水处理系统"、"废物管理"、"废物利用"]

建议规划 Jiànyì guīhuà

在美国，由规划专业人员代表某一机构、有关组织或社区针对官方规划机构所作规划表示不同或反对的建议性规划。

in opposition to planning proposal prepared by an official agency.

Aerial survey

Photographs taken from an aircraft, showing the layout of the land area below. See also REMOTE SENSING.

Aesthetic controls

In the United States, regulation by ordinance and case law of structures that violate public standards of beauty in urban and rural areas. The most common objects affected are water towers and billboards, as well as buildings inappropriate to an historic preservation district. Aesthetic controls may be exercised by community associations. See also AESTHETIC ZONING, COMMUNITY ASSOCIATION, and LAND USE CONTROLS.

Aesthetic zoning

In the United States, use of municipal ordinances to regulate the appearance of urban districts. Such zoning commonly affects such objects or establishments as signs, billboards, trail parks, and junk yards. See also AESTHETIC CONTROLS.

Age distribution

In demography, the constitution of a population

航空测量 Hángkōng cèliáng

展示地面情况的航空摄影。[参见"遥感技术"]

美观控制 Měiguān kòngzhì

在美国,指根据法令及判例法制定的规定,对在城乡地区违反公认的美观标准的构筑物(最常见者为水塔、广告牌)以及历史保护区内不适当的建筑物加以控制。美观控制可由社区协会执行。[参见"美观区划"、"社区协会"、"土地使用控制"]

美观区划 Měiguān qūhuà

在美国,指利用城市法令控制市区市容。通常对影响市容的招牌、广告牌、活动房屋集中地及杂物堆场进行区划控制。[参见"美观控制"]

年龄分布 Niánlíng fēnbù

人口统计学中各年龄段的人口构成情况。[参见

according to age groups. See also DEMOGRAPHY.

Agglomeration

A continuous group of diverse buildings in an urban area, not arranged according to any plan. The term is often used in a negative sense to mean a vast and unattractive urban region. See also BUILT-UP AREA and URBAN SPRAWL.

Agro-industrial-commercial enterprise

In China, a cooperative association of a primarily rural enterprise specializing in agricultural product processing with industrial and commercial enterprises engaging in related activities. The purpose of the joint enterprise is to meet market needs efficiently, to improve product quality, to increase product output, and to speed distribution of commodities. In the United States, the term "agrobusiness" is used for a large-scale farming or food production operation. See also INDUSTRIAL-AGRICULTRURAL ENTERPRISE.

Air rights

In the United States, the right to make use of spaces above property (such as highways, or railroad tracks), or above water for development independent of the surface below. Such rights can be purchased or leased from the landowners.

"人口统计学"]

聚集体 Jùjítǐ

在城市地区，未经规划的由各种房屋组成的连续建筑群。常含贬义，指大片不美观的城市地区。[参见"建成区"、"城市无计划扩展"]

农工商联合企业 Nónggōngshāng liánhéqǐyè

在中国，指以农产品加工为主的农村企业与有关工商企业组成的联合企业。其目的在于提高产品的产量和质量，满足市场需要，加速产品流通。在美国，"农业企业"是指大规模农场或食品加工企业。[参见"工农联合企业"]

上空使用权 Shàngkōng shǐyòngquán

在美国，指使用产业（如公路、铁路等）、水域的上部空间的权利。此种使用权可向产业所有者购买或租赁。

Alternate bid

In the United States, an amount stated in a construction bid that will be added to or deducted from the amount of the base bid if the corresponding changes in project scope, alternate materials, and/or methods of construction are accepted. See also BASE BID and BID.

Alternative technology

A substitute technology equivalent in functional value to an accepted technology. Environmentally, the term indicates a type of technology which involves minimum use of nonrenewable resources, minimum interference with the environment, regional or local self-sufficiency, and consideration for the rights and needs of individuals. See also APPROPRIATE TECHNOLOGY and INTERMEDIATE TECHNOLOGY.

Amenities

Features, items, or conveniences conducive to material comfort and a pleasant life. In an urban planning context, these may include attractive open spaces and landscape features, social and recreational facilities, and technical improvements conducive to a pleasing and agreeable urban and rural environment.

备用标价　Bèiyòng biāojià

在美国，指在工程投标书中注明当工程范围、用料与（或）施工方法有所变化时，应对基本标价作必要的相应增减的标价。［参见"基本标价"、"投标"］

替换技术　Tìhuàn jìshù

在功能上与认可的技术相当因而可以用来替代的技术。就环境而言，是指最少使用无法补充的资源，最少干扰环境，能就地取材，并符合个人权利和需要的技术。［参见"适用技术"、"中间技术"］

宜人事物　Yírén shìwù

使人们物质上舒适和生活上愉快的项目或设施。在城市规划中，可以包括优美的绿地和景观、社交和娱乐设施，以及能使城乡环境优美的技术改进。

Amortization

1. In the United States, gradual repayment of a debt with interest, usually according to a predetermined schedule of installments. The amortization schedule shows the amounts of principal and interest due at regular intervals and the unpaid balance of the loan after each payment is made. See also BALLOON MORTGAGE, FIXED-RATE MORTGAGE and MOTRGAGE.

2. In the United States, gradual phasing out of buildings and land uses that do not conform to newly instituted zoning ordinances. See also NONCONFORMING USE.

Anchor tenant

In the United States, the most important tenant in a development (for example, a shopping center), whose lease is instrumental in obtaining financing and additional tenants for the project.

Annexation

In the United States, the formal procedure for, or act of, adding a section of land to the jurisdiction of a city or town.

Apartment

A room or group of rooms in a building that is rented

1. **分期偿还** Fēnqī chánghuán

在美国,指按预定计划分期偿还债务本息,分期偿还计划还列出每一阶段所需偿还的本、息金额,以及每次付款后的贷款余额。[参见"特大尾数抵押贷款"、"固定利率抵押贷款"、"抵押贷款"]

2. **逐步淘汰** Zhúbù táotài

在美国,指对不符合新的区划法规的建筑物和土地用途,进行逐步淘汰或改变其用途。[参见"不符规定使用"]

关键租户 Guānjiàn zūhù

在美国,指一个开发区中最重要的租户。例如开发区内购物中心的租户,就是一种关键租户,它将有助于获得建设资金并吸引其他租户。

合并土地 Hébìng tǔdì

在美国,指将部分土地正式并入城镇管辖范围。

公寓 Gōngyù

建筑物内租予住户的一间或一套居住单元,带有

or leased to a tenant and constitutes a self-contained dwelling unit. Called a flat in Great Britain. See also GARDEN APARTMENT and TENEMENT BUILDING.

Apartment house

A building of two or more floors containing separate apartments, that is, apartments with their own entrances, approached either from a common staircase, from a common staircase, from a common access balcony leading from a common staircase, or from a common elevator. The British equivalent is a block of flats. In China, such a building is called, literally, "multiunit housing." See also MULTIFAMILY HOUSING and TENEMENT BUILDING.

Application for payment

In the United States, a contractor's written request for payment of the amount due for completed portions of the work and, if the contract so provides, for materials delivered and suitably stored prior to their use.

Apportionment

Division and distribution, especially of land and funds, by a public authority according to a plan or set of criteria.

独用厨房、浴室等设备。在英国,称为套房。[参见"花园公寓"、"经济公寓"]

公寓楼 Gōngyùlóu

包括多套公寓的两层或两层以上的住宅楼。每套公寓都有独立的出入口,由共用楼梯或共用外廊出入。在英国称为套房楼,在中国则称为"单元式住宅"。[参见"多户住房"、"经济公寓"]

付款申请书 Fùkuǎn shēnqǐngshū

在美国,指承包人对已完成部分的工程提出的付款申请书,如在合同中已有规定,也可在备用材料运到工地并存放妥善后提出。

土地或资金分配 Tǔdì huò zījīn fēnpèi

有官方权威机构根据规则或准则划分并分配土地或资金。

Appraisal

In the United States, evaluation or estimation, preferably by a qualified professional appraiser, of the value, cost, of utility of land or property. See also ASSESSED VALUE and HIGHEST AND BEST USE.

Appreciation

An increase in the value of currency, goods or property, usually attributable to changes in economic circumstances. An increase in the value of property usually results from a combination of inflation, higher demand, and increase accessibility, as opposed to increase in value attributable to improvements of the property. See also DEPRECIATION and SHARED APPRECIATION MORTGAGE.

Appropriate technology

A level of technology in keeping with the available resources, e.g., the size of the labor force and availability of capital, and the technical state-of-the-art. See also ALTERNATIVE TECHNOLOGY.

Appropriation

Authorization by an official person or group permitting use of a fixed sum of money, usually for a one-year period or a fixed number of years, for a specific purpose or project.

估价 Gūjià

在美国,指由合格的专职估价人员对土地或产业的价值、价格或效用作出估价。[参见"房地产课税价值"、"最佳用途"]

增值 Zēngzhí

由于经济情况变化引起的货币、货物或产业价值的增加。房地产的增值往往是通货膨胀、需求增加以及交通条件改善的结果,而并非改进房地产本身所引起的价值增加。[参见"折旧、贬值","分享增值抵押贷款"]

适用技术 Shìyòng jìshù

与现有资源,如劳动力、生产规模,可能投入的资金和当期技术水平相适应的技术。[参见"替换技术"]

拨款 Bōkuǎn

由官方人员或机构批准供某一特定目的或工程使用的一定数额的款项。通常以一年为期或在规定年限内使用。

Approved equal substitution

In the United States, material, equipment, or methods, approved by the architect or engineer for use in a project because they are considered equivalent in essential attributes to the material, equipment, or method specified in the contract documents.

Aquifer

An underground bed or water-permeable stratum of rock, gravel, or sand that collects water and serves as the source of springs and wells.

Arbitration

In the United States, the hearing and settlement of disputes, usually about contracts, by an arbitrator or arbitrators chosen by the disputants or appointed by a governmental or judicial authority. Arbitration is used as an alternative to litigation.

Architect

Official designation for a person or organization professionally qualified and duly licensed to perform architectural services. In the United States, basic architectural services, as contractually defined, consist of schematic design, design development, preparation of construction documents, participation in the bidding or negotiation process, and construction document administration. Comprehensive services

准许代用品 Zhǔnxǔ dàiyòngpǐn

在美国，指由建筑师或工程师批准，认为与合同文件中规定的材料、设备或方法在基本属性上相等，可以在工程中采用的代用品。

蓄水层 Xùshuǐcéng

能积水并为泉、井水源的岩石、砾石或沙的渗水地层。

仲裁 Zhòngcái

在美国，合同双方发生争执时，由争执双方选定或由政府司法部门制定一个或若干个公断人，听取论据并作出裁决，以替代公诉。

建筑师、建筑师事务所 Jiànzhùshī、Jiànzhùshī shìwùsuǒ

由政府确认合格，领有执照，准许从事建筑业务的个人或机构，常为法定职称。在美国，其基本任务包括承担方案设计、技术设计、施工图设计、参加投标及谈判、管理施工文件等。其业务范围还可扩大到承担可行性研究、施工管理及专业性咨询等。在中国，建筑设计业务通常由设计院承担。[参见"建筑师工程师事务所"]

extend beyond traditional services to feasibility studies, construction management, and special consulting services. In China, most architectural services are performed by design institutes. See also ARCHITECTURAL ENGINEERING FIRM.

Architect's approval

In the United States, architect's or engineer's written acknowledgement that any phase of building construction is acceptable, or that a contractor's request of claim is valid.

Architectural engineering firm

A commercial organization offering the professional service of architects and engineers on a contract basis. See also ARCHITECT and DESIGN-BUILD PROCESS.

Architectural services

See ARCHITECT.

Architecture

The art or practice of designing buildings in accordance with principles determined by aesthetic and practical or material considerations. See also ENVIRONMENTAL DESIGN.

建筑师认可证明 Jiànzhùshī rènkězhèngmíng

在美国，指建筑师所开具的书面证书，认可某一阶段的施工工程可以接受，或承包人的付款申请正当。

建筑师工程师事务所 Jiànzhùshī gōngchéngshī shìwùsuǒ

建筑师与工程师以承包的方式提供专业服务的营利性机构。[参见"建筑师、建筑师事务所"，"设计兼施工"]

建筑业务 Jiànzhùyèwù

[见"建筑师、建筑师事务所"]

建筑学 Jiànzhùxué

按照美学观点与实际需要与物质条件的许可所确定的原则，设计建筑物的技艺或实践。[参见"环境设计"]

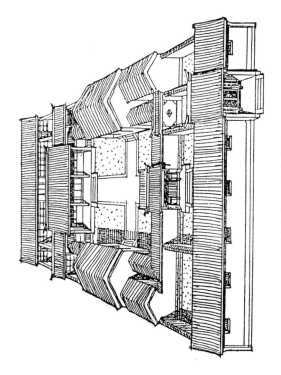

中国北京四合院式居民
Courtyard house in Beijing

1. 堂/厨　Living/Kitchen
2. 卧室　Bedroom
3. 院子　Courtyard
4. 渗井　Well
5. 厕所　Toilet

中国河南省窑洞式民居
Cave dwellings in Henan Province

Area method

In the United States, a technique for estimating construction costs of a building. The architectural area, i.e., the sum of the adjusted areas of the building's floors, is multiplied by the current cost per unit of area.

Area take-off

In the United States, a rough estimate of the construction costs for specific types of areas, within a proposed building, measured as cost per square foot for each area type. The area take-off is usually prepared during the schematic design phase. See also SCHEMATIC DESIGN.

ARM

See ADJUSTABLE RATE MORTGAGE.

Arterial road

A main or trunk road in the street system of an urban area.

Artery

A general term for the main line of railway, airline, highway, and water-way transportation.

Assessed value

In the United States, the value of land and buildings,

面积估价法 Miànjī gūjiàfǎ

在美国，指用建筑面积估算建筑物工程承包的方法。即建筑物各楼层面积之和乘以当时的单位面积成本。

粗略估算 Cūlüègūsuàn

在美国，根据拟建工程中各不同类型部分面积的单位成本，粗略地估算拟建工程成本，通常用于方案设计阶段。[参见"方案设计"]

可变利率抵押贷款（缩写） Kěbiàn lìlǜ dǐyā dàikuǎn (suōxiě)

[见"可变利率抵押贷款"]

主干道 Zhǔgàndào

城市道路网中的骨干道路。

干线 Gànxiàn

铁路、航空、公路、水路交通等骨干线路的通称。

房地产课税价值 Fángdìchǎn kèshuì jiàzhí

在美国，由地方当局为征税目的而定的房地产价

often represented as a percentage of the property's market value, set by a local government for tax purpose. See also APPRAISAL and CADASTRAL SURVEY.

Assets

In the United States, all of the holdings of a company or government with a money value as listed on a balance sheet. They are classified as tangible fixed assets such as machinery, intengible assets such as goodwill, and current assets such as stock and cash. See also LIABILITIES.

Assignment of title

In the United States, legal transfer of ownership of land or property from one individual to another by recording the transaction on the official record of ownership. Contract or option rights and leases may also be assigned by one party to another. See also LAND ACQUISITION and TITLE.

Assisted housing

In the United States, privately owned dwelling units for needy households constructed or rehabilitated with the assistance of government grants or low-interest loans. Alternatively, subsidies may be granted to housing suppliers willing to reduce rents for low-

值。通常以该房地产市场价值的百分数表示。[参见"估价"、"地籍册"]

资产 Zīchǎn

在美国,企业或政府列入其资产负债表中的一切具有货币价值的财产,分为:有形固定资产,如机器;无形资产,如信誉;流动资产,如股份与现金。[参见"负债"]

过户 Guòhù

在美国,指将土地或地产所有权合法转让给他人,并正式登记备案。合同、租约或买卖权亦可转让。[参见"征地"、"产权证"]

公助住房 Gōngzhù zhùfáng

在美国,指由政府资助或以低息贷款建造或改建的供需房户居住的私有住房,或者由政府对于愿对低收入住户减低租金的房地产主给予津贴。在美国住房市场中,公助住房的比例很小(不足5%)。在一般用语中,公助住房包括公共住房,

income tenants. Assisted housing represents a very small proportion of the American housing market (less than 5 percent). In common usage, assisted housing may include public housing and may also be called subsidized housing or low-and-moderate-income housing. See also HOUSING ALLOWANCE, HOUSING MANAGEMENT, and PUBLIC HOUSING.

Atrium architecture

A contemporary building style in which all or part of the interior space of a building is left without barriers to a height of several stories and is landscaped with shrubs and trees. In small structures, the atrium may be open at the top and walled with glass. In high-rise buildings, the atrium area often serves as a lobby for the building or as a promenade lined with shops and restaurants.

Attached house

A house that shares common walls with the adjacent houses on either side, i.e., a townhouse or rowhouse. See also DETACHED HOUSE, DUPLEX HOUSE, QUADRUPLEX HOUSE, ROWHOUSE, TOWNHOUSE, and TRIPLEX HOUSE.

Autonomous region

In China, an administrative area of a minority

又称"补贴住房","中低收入住房"。[参见"住房津贴"、"住房管理"、"公共住房"]

内庭式建筑 Nèitíngshì jiànzhù

一种现代的建筑风格,建筑物内部有高达数层无间隔、植以树木的空间,称为内庭。在小型建筑中,内庭可用玻璃做墙,上无屋顶;在高层建筑中,内庭常用作门厅或休息场所,四周设店铺、餐厅等。

毗连住宅 Pílián zhùzhái

与相邻住宅共有分界墙的住宅,即联立式住宅或市镇联立式住宅。[参见"独立式住宅"、"二联式住宅"、"四联式住宅"、"联立式住宅"、"市镇联立式住宅"、"三联式住宅"]

自治区 Zìzhìqū

中国由少数民族自治的地方,相当于省一级的行

nationality, corresponding to a province, such as the Xizang (Tibet) Autonomous Region where most of the population is Tibetan, and the Guangxi Zhuang Autonomous Region where the population consists mostly of the Zhuang people. The Constitution of China stipulates that the administrative hierarch is 1) autonomous region, 2) autonomous prefecture, and 3) autonomous county. See also MUNICIPALITY.

Average floor area per unit

In China, the total floor area of the dwelling units in a housing project divided by the total number of units. See also DWELLING SIZE.

Average living floor area per capita

In China, the total living floor area in a housing project divided by the total number of persons living there.

Average number of persons per household

In China, the total number of registered residents divided by the number of households in a given area.

AWT

See ADVANCED WASTEWATER TREATMENT.

政区域单位。例如以藏族为主的西藏自治区，以壮族为主的广西壮族自治区。中国宪法规定自治地方的行政建制为三级：自治区、自治州和自治县。[参见"市"]

平均每套建筑面积 Píngjūn měitào jiànzhù miànjī

在中国，指各套建筑面积之和除以总套数。[参见"户型"]

平均每人居住面积 Píngjūn měirén jūzhù miànjī

在中国，指各套住宅居住面积之和除以各套住宅中居住的总人数。

平均每户人口 Píngjūn měihù rénkǒu

在中国，指某一地区内平均每户有正式户口的人口数。

污水深度处理（缩写） Wūshuǐ shēndù chǔlǐ (suōxiě)

[见"污水深度处理"]

B

Balloon mortgage

In the United States, a mortgage arranged with two separate phases of payment, A schedule for amortization covers a specified part of the total loan. The remainder of the loan is paid off in a single lump sum payment at the termination of the contract. See also AMORTIZATION and MORTGAGE.

Barrier-free environment

In the United States, a living environment in which all of the facilities and services, including buildings, gates, doorways, parking lots, transportation, etc., are constructed to permit access and mobility of handicapped individuals, especially those in wheelchairs.

Base bid

In the United States, the amount of money stated in the bid as the sum for which the bidder offers to perform the work, not including that work for which alternate bids are submitted. See also ALTERNATE BID, BID, and CONSTRUCTION COST.

Base rent

In the United States, the minimum fixed rent for a commercial property.

Baseline data

Specific information items gathered in a given time

特大尾数抵押贷款 Tèdà wěishù dǐyā dàikuǎn

在美国,一种分两阶段偿付的贷款方式。一部分按计划分期偿付,剩余部分在合同到期时一次付清。[参见"分期偿还"、"抵押贷款"]

无障碍环境 Wúzhàng'ài huánjìng

在美国,指一切设施和设备(包括建筑物、大门、门口、停车场、运输工具等)的构造均便于残疾人(特别是坐轮椅的人)出入和活动的生活环境。

基本标价 Jīběn biāojià

在美国,投标人在投标书中所列出的完成工程的总金额,不包括备用标价的工程项目。[参见"备用标价"、"投标"、"建筑成本"]

基本租金 Jīběn zūjīn

在美国,指商业房地产最低的固定租金。

基准数据 Jīzhǔn shùjù

在指定时期内与/或在正常情况下收集到的数据,

period and/or under normal conditions to provide a standard of comparison for future survey or experimental data.

Beautification

Measures such as tree planting or general cleaning of exterior surfaces of buildings, taken to enhance the general attractiveness of the urban environment.

Bedroom suburb

In the United States, outlying residential area from which residents commute to work centers of the city. Also called a dormitory suburb.

Beltway

In the United States, a multilane, limited-access highway built encircling a major urban area, facilitating rapid travel from one suburban area to another and permitting long distance traffic to bypass the city entirely. See also EXPRESSWAY.

Berm

In the United States, a linear embankment constructed to direct the flow of surface drainage, to provide a visual barrier, and/or to help reduce the traffic noise in adjacent residential areas. The berm is usually covered with grass to resist erosion and is often planted with large shrubs.

用来作为日后调查或实验数据的对比标准。

美化 Měihuà

改进城市环境或市容的一种措施,如植树、对建筑物表面的清理等。

卧城郊区 Wòchéng jiāoqū

在美国,指市区外围的居住区。居民每日往返于居住区与市内工作地点之间。又称"卧城区"。

环城公路 Huánchéng gōnglù

在美国,指在重要城市地区周围,限制支路通入的多车道公路,以便从一个郊区迅速到达另一个郊区,并完全避免长途交通穿过市区。[参见"高速公路"]

路肩 Lùjiān

在美国,指公路两旁的护堤。这种护堤便于路面排水,既可作为视觉屏障,又有助于降低对附近居住区的交通噪声。路肩上常植以草被或大型灌木以防止水土流失。

Bid

In the United States, a formal offer by a contractor or commercial organization to complete work or to provide services at the price specified in a bid schedule by a prescribed time. In the construction industry, the offer submitted by the bidder in the required manner may include provision of all labor, equipment and material, and performance of the specified work. See also ACCEPTED BID, ALTERNATE BID, BASE BID, BID ABSTRACT, CONSTRUCTION BUDGET, and CONTRACT AWARD.

Bid abstract

In the United States, a summary of unit prices compiled by the owner or his representative for a given job. An abstract may list bidders and their bid prices on a given project, usually broken down by items in the project. Such a summary facilitates selection of the contractor to be awarded the job on the basis of the low bid. See also BID and CONTRACT AWARD.

Bid bond

In the United States, a form of bid security with the bidder as principal and with a surety, i.e., a guarantor, to ensure that the bidder will enter into a contract by a given date. See also BOND.

投标 Tóubiāo

在美国,由承包商或商业组织提出的愿在预定时期内按一定价格完成工程或提供服务的正式报价。在建筑业中,由投标人按指定方式所提出的报价,包括为指定工程提供全部人工、设备、材料,并为工程施工。[参见"中选标"、"备用标价"、"基本标价"、"标价总表"、"工程预算"、"发包"]

标价总表 Biāojià zǒngbiǎo

在美国,指由业主或其代理人为某一工程准备的单价汇总表,常按工程项目分别列出各投标人及其投标价格,以便根据低价的原则选择承包人。[参见"投标"、"发包"]

投标保单 Tóubiāo bǎodān

在美国,由投标人及保证人所具有的保证书,保证投标人在规定日期内履行合同。[参见"保证书"]

Bid price

1. In the United States, the amount that a buyer offers to pay for a product or property.
2. In the United States, the amount that a contractor or commercial organization proposes for completion of work.

Bidding documents

In the United States, the written descriptions of bidding requirements, for example, the invitation to bid and the instructions to bidders; sample bidding and contract forms that define the requirements for submitting bids; and addenda such as formal responses to questions from potential bidders. See also ADDENDUM, DEPOSIT FOR BIDDING DOCUMENTS, and REQUEST FOR PROPOSAL.

Bidding requirements

See BIDDING DOCUMENTS.

Bikeway

A path specially constructed for bicycle traffic. Bikeways may be separate lanes along the sides of streets, or in the United States, may be located in open spaces or parks, primarily for recreational purposes.

1. **出价** Chūjià

在美国,指买方对于购置某一产品或产业所愿付的金额。

2. **投标价格** Tóubiāo jiàgé

在美国,指承包人或商业组织为完成工程所需而提出的金额。

招标文件 Zhāobiāo wénjiàn

在美国,指说明投标条件的文件,包括投标邀请书、投标须知;说明投标要求的标单格式及合同格式的样本;补充文件,如对于参加投标者所提问题的正式答复等。[参见"补充文件"、"投标文件押金"、"招标通知"]

投标条件 Tóubiāo tiáojiàn

[见"招标文件"]

自行车道 Zìxíngchēdào

专为自行车通行的道路,可为在街道两旁的专用车道。在美国,绿地或园林内也常设有供娱乐用的自行车道。

Bill of quantities

See QUANTITY SURVEY.

Bill of sale

In the United States, a legal document registering the transfer of ownership of property or goods from one party to anther under given terms at a specified prices.

Blanket mortgage

A lien on more than one parcel or unit of land, frequently incurred by subdividers or developers who have purchased a single tract of land for the purpose of dividing it into smaller parcels for sale or development. Also called a blanket trust deed.

Blanket trust deed

See BLANKET MORTGAGE.

Blight

In the United States, term used to characterize the deteriorating condition of specific buildings or neighborhoods within a given city. See also SLUM and URBAN REDEVELOPENT.

Block grant

In the United States, federal funds provided to local governments to help finance local services such as roads, education, and general urban development.

数量明细表 Shùliàng míngxìbiǎo

［见"工程用料与设备清单"］

卖契 Màiqì

在美国，指载明产业或货物的所有权，以一定价格在指定条件下由一方转让给另一方的法律性文件。

揽总抵押 Lǎnzǒng dǐyā

以许多地块为抵押品的留置权，通常是由于土地再划分者或开发者购得整块土地后划成小块出售或开发所形成。又称为揽总抵押信托书。

揽总抵押信托书 Lǎnzǒng dǐyā xìntuōshū

［见"揽总抵押"］

衰落 Shuāiluò

在美国，指城市中某些建筑物或住宅区的衰退情况。［参见"贫民窟"、"城市改造"］

一揽子赠款 Yīlǎnzi zèngkuǎn

在美国，由联邦政府向地方政府提供的资金，用来资助当地的建设，例如修道路、办教育、进行城市的一般建设等。一揽子赠款是分享财政收入

Block grants are a type of revenue sharing in which funds are distributed on a formula basis rather than competitively. See also as COMPETITIVE FUNDING, ENTITLEMENT CITY, FORMULA FUNDING, and GRANT.

Block of flats

See APARTMENT HOUSE.

Board of zoning appeals

In the United States, a quasi-independent local body appointed by the mayor or serving as a committee of the city council. The board is charged with hearing appeals of decisions by zoning enforcement officials, granting zoning variances, and making decisions on applications for special exceptions to zoning or for special use permits. See also CITY PLANNING COMMISSION, SPECIAL USE PERMIT, and ZONING VARIANCE.

Boardinghouse

In the United States, a private residence whose owners supply moderately priced meals, and often rooms, to individuals.

Bond

1. In the United States, an agreement guaranteeing that a contracting party will perform certain work

的方式之一，其资金按公式拨款而不采取竞争方式。[参见"竞争性拨款"、"受资助城市"、"按公式拨款"、"赠款"]

套楼房 Tàolóufáng

[见"公寓楼"]

区划申诉委员会 Qūhuà shēnsù wěiyuánhuì

在美国，由市长任命或作为市议会的委员会之一的半独立性地方机构，负责听取对于区划执行官的决定的意见，准许区划宽限，对于免受区划限制或特殊用途的申请作出决定。[参见"城市规划委员会"、"专用许可证"、"区划特许证"]

供膳宿的私家住宅 Gōngshànsùde sījiā zhùzhái

在美国，指由房主供应一般价格的膳食及住宿的私人住宅。

1. 保证书 Bǎozhèngshū

在美国，指订立合同的一方保证在规定日期内完

or fulfill certain obligations by a specified date and that in the event of nonperformance or default the other contracting party will be indemnified. See also BID BOUND, MAINTENANCE BOND, and PERFORMANCE BOND.

2. In the United States, a certificate issued by a government or corporation guaranteeing the purchaser a fixed rate of interest. Unlike the shotterm loan, the bond does not require repayment of principal until the bond is retired. Government bonds backed by the fees charged to users of a structure or service financed by the sale of the bonds are referred to as revenue bonds. The purpose for which a bond is issued is frequently indicated in the name, e.g., a water and sewer bond. See also INDUSTRIAL REVENUE BOND and MORTGAGE REVENUE BOND.

Bonus and penalty clause

In the United States, a provision in a contract stipulating that the contractor will be paid a bonus amount for completing work prior to the specified final date or will be charged a penalty for failure to complete work by that date.

Bonus incentives for development

In the United States, special advantages granted

成某项工作或某项义务的协议书,如承包的一方未能完成协议的规定或违约,则订立合同的另一方将得到赔偿。[参见"投标保单"、"完工保证书"、"维修保证书"、"履行合同保单"]

2. 债券　Zhàiquàn

在美国,指由政府或企业发行的一种证券,保证购券人可得到固定利率的利息,这种债券和短期借款不同,在债券期满前并不要求偿还本金。由政府发行,通过向使用者取费回收的债券,或为某些服务项目提供资金而发行的债券都称为收益债券。债券一般按项目定名,例如"上下水债券"。[参见"工业债券"、"抵押受益券"]

奖罚条款　Jiǎngfá tiáokuǎn

在美国,指合同中的一项条款,规定承包人如在完工日期以前完成施工任务,可得到奖金,如不能按期完工则将处以罚款。

开发奖励　Kāifā jiǎnglì

在美国,由社区授予城市建设投资人的一种特权,

to a developer by the community in exchange for a benefit that the community may not be able to require, for example, permission to erect a larger-than-usual building in exchange for reasonably priced retail rental space on the lower floors. See also INCENTIVE ZONING.

"Borrowed view"

In the classic garden design of China, the view presented to the visitor was enlivened by incorporating scenes from the outside world of nature into the natural world of the garden itself. This concept can be applied similarly in urban design to enhance the visual impression within the city by drawing distant views into the urban setting. See also LANDSCAPE ARCHITECTURE, "SCENIC FOCAL POINT," and "YIJING."

"Bottom line"

In everyday American speech, the most important factor or result in a process, project, or commercial undertaking. The term actually refers to the net income line on an income statement and the thus to the net profit of a company, unit, or product. While the term has its origins in accounting, "bottom line" may just as easily refer to a nonmonetary result, e.g., an important personnel change.

以换取社区无力获得的利益。开发者则将建筑物的下面几层以合理价格租给零售商店,作为交换条件。[参见"鼓励性区划"]

借景 Jièjǐng

中国古典园林的一种设计艺术,即将外部景物纳入园内,使之与园内景色融为一体。这种手法亦可用作城市规划设计,即将远方景色收入城市环境之中。[参见"风景建筑学"、"对景"、"意境"]

净收益栏 Jìngshōuyìlán

原指会计学上表示净收益的专栏项目,即一个公司、单位或产品的净利润。在美国的日常用语中则指一个过程、计划或商业业务中最重要的因素或结果。此词有时也可引申为非经济性的结果,如因某些事态所导致的重要人事更动等。

Boundary survey

See LAND SERVEY.

"Brainstorming"

An informal, often intensive, group discussion to solve a difficult problem or to develop new approaches to an issue. Those involved usually meet under a tight deadline. Ideas evolving from the discussion are closely examined, with those surviving being expanded until consensus is reached on a particular solution or set of options. See also CONCEPT DEVELOPMENT.

Break-even point

The point at which sales or production of a product are sufficient to cover related costs without either monetary gains or losses, In the American real estate sector, the break-even point is the figure at which income derived from rentals of commercial or residential property equals all the required expenses for the property and debt service.

Buffer zone

A section of land that is intended to separate areas with incompatible uses, e.g., industrially zoned land from residentially zoned land, or in China, warehouse and transportation facilities from cultural districts. The purpose of such zones is to limit the effects

地界测量　Dìjiè cèliáng

［见"土地测量"］

绞脑汁会　Jiǎonǎozhīhuì

为解决难题或探索新途径而采取的一种非正式的、讨论激烈的集体献计方式。通常为与会者规定时限，会上发表的各种意见经过严格筛选，并对好的意见加以发展，直至获得一项或数项决议。［参见"概念发展"］

收支平衡点　Shōuzhī pínghéngdiǎn

当产品的销售或产量恰好抵消各项成本，既无利润又无损失时，即为收支平衡点。美国房地产业的收支平衡点是指商业或居住用房地产的租金收入与全部开支及应偿付债务相等时的租金额。

防护带　Fánghùdài

用以分隔不同用途的区域之间的土地。例如，划作工业用的土地与居住用的土地之间，仓库与交通设施区以及文教区之间。防护带的作用在于减少噪声、不雅观、烟尘、爆炸、火灾、臭气或有毒气体，除绿化以外不作其他处理。在中国，不

of noise, visual ugliness, smoke, explosions, fires, odors, or poisonous gases. Buffer zones are usually unimproved except for landscaping, and in China, no construction of any kind is allowed in the zone.

Buildable area

In the United Stated, the area remaining for construction after minimum yard and open space requirements have been met.

Building alteration

Construction work to revise prescribed elements of an existing structure, as distinct from additions to an existing structure. See also ADDITION and HOME IMPROVEMENT.

Building area

In the United States, the surface area taken up by all principal and assessory buildings on a lot, measured from the outside of the exterior walls on the ground floor, and not including uncovered entrance platforms, terraces, and steps.

Building codes and standards

1. In the United Stated, a body of ordinances and regulations adopted by a state or local government, controlling the design, construction, materials, alteration, and occupancy of any building or

允许在这部分土地上建造任何建筑物。

可建面积 Kějiàn miànjī

在美国,指除留作院落和绿地用的最低限度的空地外,可供建筑用的土地。

改建工程 Gǎijiàn gōngchéng

对现有建筑中的某一部分进行改建的工程。不同于加建工程。[参见"加建工程"、"住房改善"]

基底面积 Jīdǐmiànjī

在美国,指基地上由主要建筑物和附属建筑物所占的土地面积,自底层外墙面计算,不包括无顶的入口平台、露台和台阶。

建筑法规与标准 Jiànzhù fǎguī yǔ biāozhǔn

1. 在美国,指州政府或地方政府采用的一系列法令和规范,用以管理所辖区内任何建筑物或构筑物的设计、施工、材料、改建和使用。建筑法规可由警方实施,以保护公众健康、安全和福利。

structure within its jurisdiction. The building code is enforceable by police powers for the protection of public health, safety, and welfare. The code often includes technical standards for all critical building installations, including electrical, mechanical, plumbing, and heating systems. Thus, codes can be viewed as a collection of standards necessary to ensure a basic minimum level for design and construction within a specific area.

In practice, the system of codes is complex. Within the United States, there are about 12,000 building code jurisdictions, the large majority of which adopt or adapt one of three model codes developed by three different nongovernmental bodies: the Basic Build Code (BBC); the Southern Standard Building Code (the Southern); or the Uniform Build Code (UBC). States and local governments can take one of four approaches: 1) they can adopt a model code in its entirety and give it the force of law in their jurisdiction; 2) they can adapt a model code to their particular area; 3) they can develop their own code, as New York City and Chicago have done; or 4) they can have no build code. [If a local government have no build code, it often adopts technical standards for one or more elements, e.g., the National Electrical Code (NEC) for electrical systems.]

法规中一般包括建筑物重要设备（包括电力、机械、给排水及采暖各系统）的技术标准。因此，建筑法规也可以说是在指定地区内保证设计和施工达到最基本水平所必需的标准的总和。

在实践中，法规的系统颇为复杂，在美国国内约有12000个建筑法规辖区，其中大多数采用或套用三个非政府机构所编制的示范法规中的一种：①《基本建筑法规》（BBC）；②《南方标准法规》；③《统一建筑法规》（UBC）。州政府和地方政府可采用以下四种措施之一：①全盘采用一种示范法规，使之在辖区内具有法律效力；②将一种示范法规加以改编，使之适合其辖区的特殊情况；③自行编制本地区的法规，例如纽约市及芝加哥制定的建筑法规；④地方政府不自行编制建筑法规，而采用有关的技术标准，例如在电力系统中采用《国家电力法规》（NEC）。

Although about half the states have adopted statewide building codes, they have generally left interpretation and enforcement to local governments. The construction standards for manufactured housing (mobile house), developed by the U.S. Department of Housing and Urban Department (HUD), constituted the only national building code. HUD also publishes the Minimum Property Standard (MPS), but its use is generally restricted to construction of federally assisted housing, about five percent of the total housing stock. However, the MPS is currently being phased out in favor of a new model residential building code for one-and two-family dwellings developed by the Council of American Building Officials (CABO), another nongovernmental organization. See also BULK REGULATIONS, CODE ENFORCEMENT, CONSTRUCTION SPECIFICATIONS INSTITUTE, MODEL CODE, PERFORMANCE CODE, SPERCIFICATION CODE, SPECIFICATIONS, and SUNLIGHT ORDINANCE.

2. In China, building and construction laws are enacted by the Standing Committee of the National People's Congress, while ordinances are approved by the State Council. Both forms are applicable nationwide. Since the early 1970's, efforts have been

现在美国虽然约有半数的州已采用全州性的建筑法规,但解释权和执行权一般属于地方政府,由美国住房与城市发展部(HUD)所制订的活动房屋构造标准是唯一的全国性建筑法规,该部所颁布的《最低房产标准》(MPS),仅适用于建造受联邦政府补助的住房,约占现有住房总量的5%。但目前,《最低房产标准》已逐渐被淘汰,而代之以另一非政府组织——美国建筑官员委员会(CABO),为1—2户的住宅所制定的新的示范住房法规。[参见"体积规定"、"执行法规"、"施工规范研究院"、"样板法规"、"功能法规"、"规格法规"、"说明书"、"日照标准"]

2. 在中国,建筑与施工方面的法由人民代表大会常务委员会发布。建筑法令,由国务院批准,二者均在全国范围内执行。自20世纪70年代以来,建筑及住房方面的立法工作,如结构工程、建筑产品安全等方面,已取得一定成果。有关城市规

underway to codify laws relating to construction and housing, with some success in the areas of engineering, products, and building safety. The first administrative ordinance on urban planning was issued by the State Council in Spring of 1984. Codification efforts in the areas of architectural design and housing equipment have yet to be completed.

Codes (regulations), which are usually also promulgated at a national level, are technical documents providing mandatory direction for design and construction work. Most codes in China are performance codes. The codes are supplemented by a series of specifications which give detailed performance criteria for certain types of engineering. These specifications are generally approved by ministries or local governments. Standards, as determined by the State Bureau of Standardization, may be divided into four levels: 1) national standards that are of great significance to the national economy and technical development and therefore must be uniform throughout the country; 2) specialized or ministerial standards that are unified within a specialty or ministry; 3) local standards that are neither national nor ministerial (specialized) standards and that are approved by local authorities;

划的第一部法令1984年春由国务院颁发。在建筑设计及住房设备方面的立法，尚待继续进行。

建筑规范一般均为国家级，为指导设计和施工的强制性技术文件，大多数为功能规范。规程是根据规范制定的一系列说明具体技术要求的文件，由有关部门或地方政府批准。标准由国家标准局确定，可分为四级。①国家标准：在全国范围内统一执行，对国民经济和技术发展有重大关系的标准。②部门或专业标准：由专业部门或专业标准化组织批准发布，在某部门或专业范围内统一执行的标准。③地方标准：除国家标准和部门标准外，由某一地方主管部门批准发布，在该地方范围内统一执行的标准。④企业标准：一般为产品标准。由企业单位或其上级有关机构批准发布，在该企业内统一执行的标准。

and 4) company standards that mainly apply to products and that are approved by the upper-level administrative departments of companies.

Building complex

In the United States, a group of buildings, usually constructed as part of the same project and closely related in design and/or function.

Building density

As used in the United States, the ratio between the built-on area and the total area of the land on which a building is situated.

Building efficiency

In the United States, the percentage ratio of the net leasable area of a building to the total area of that building. See also LEASABLE AREA.

Building height

The size of a building in the vertical dimension, as measured in stories or linear feet/meters. Building height is often regulated in urban areas. According to U.S. standards, a low-rise building usually has one or two stories; a mid-rise building, three to seven stories; a high-rise building, more than seven stories. According to Chinese specifications for housing, a low-rise building has one to three stories;

建筑群 Jiànzhùqún

在美国一般指属于同一建设项目,在设计与 / 或功能上密切相关的一组建筑。

建筑密度 Jiànzhù mìdù

在美国,指建筑占地面积与基地面积之比。

可出租面积比 Kěchūzū miànjībǐ

在美国,指建筑物内可供出租的建筑面积与全部建筑面积之比。[参见"可出租面积"]

建筑高度 Jiànzhù gāodù

建筑物竖向的量度,以层数或尺寸(英尺或米)计。市区的建筑高度一般有规定。在美国,常以一、二层为低层,三至七层为中层,超过七层者为高层。在中国,根据住宅建筑设计规范,一至三层为低层,四至六层为多层,七至九层为中高层,十至三十层为高层。[参见"建筑高度区划"]

a multistory building, four to six stories; a medium high-rise building, seven to nine stories; and a high-rise building, ten to thirty stories. See also HIGHT ZONING.

Building inspector

In the United States, a city official charged with enforcing building codes and zoning regulations through examination of building plans and construction of structures and through the authority to issue building permits required for building construction and use permits necessary for occupancy of a finished building. See also CODE ENFORCEMENT, SITE INSPECTION, and ZONING ADMINISTRATION.

Building law

See BUILDING CODES AND STANDARDS.

Building line

A line beyond which construction of the facade of a building is prohibited according to local building regulations.

Building material testing

Assessment of specific types of materials used in construction according to performance criteria, standards, and procedures established by a

建筑监察员 Jiànzhù jiāncháyuán

美国市政府的官员之一,其职责为通过对建筑设计及施工的审查,监督建筑法规及区划规定的实施,并有权颁发施工执照及已竣工房屋的使用许可证。[参见"执行法规"、"现场监督"、"区划执行官"]

建筑法 Jiànzhùfǎ

[见"建筑法规与标准"]

建筑红线 Jiànzhù hóngxiàn

按照地方建筑规范的规定,建筑物的外立面所不能超出的界线。

建筑材料检验 Jiànzhù cáiliào jiǎnyàn

根据政府机构(如美国国家标准局)或公认的标准委员会、工业协会、专业团体或私营试验机构,如美国试验与材料学会(ASTM),所规定的功能

government agency such as the U.S. Bureau of Standards, or by consensus standards committees, industry associations, professional groups, or private testing organizations. The American Society for Testing and Materials (ASTM) publishes many of these tests. See also ACCREDITATION, CERTIFICATION OF MATERIALS AND PRODUCTS, PERFORMANCE CODE, and QUALITY CONTROL.

Building ordinance

See BUILDING CODES AND STANDARDS.

Building orientation

Situation of land or a building with reference to a given direction, which determines the effects of weather, sunlight, and environment.

Building permit

An official document issued by the appropriate government agency authorizing construction of a project in accordance with approved drawings and specifications. In China, a building permit is issued only after land acquisition, the layout plan, and the schematic or preliminary design have been approved by the urban planning administration bureau. See also ZONING ADMINISTRATOR.

准则、标准和步骤,对于某种类型的建筑材料作出评定。[参见"确认"、"材料与制品鉴定"、"功能法规"、"质量控制"]

建筑法令 Jiànzhù fǎlìng

[见"建筑法规与标准"]

朝向 Cháoxiàng

指一块土地或建筑物在基本方位上的方向。朝向对气候、日照及环境的影响起决定作用。

施工执照 Shīgōng zhízhào

由政府有关部门签发的一种工程施工许可证。允许施工单位根据批准的图纸及说明书进行工程施工。在中国,施工执照仅在城市规划管理局已经拨地,并批准总平面布置图和设计方案或初步设计后,方可签发。[参见"区划行政官"]

Building physics

The body of knowledge about physical principles that affect the structure and design of buildings, for example, acoustics.

Building regulations

See BUILDING CODES AND STANDARDS.

Building society

See SAVINGS AND LOAN ASSOCIATION.

Building standards

See BUILDING CODES AND STANDARDS.

Building trade

Any one of the skill groups and subgroups used in the construction industry.

Built environment

1. An area in which buildings have been constructed.
2. The character of a heavily built urban area, as contrasted to the natural environment.

Built-up area

A region that has been extensively developed and is already occupied by structures, outbuildings, and service areas. In modern usage, the term is often synonymous with "urban" in its physical sense. See

建筑物理 Jiànzhù wùlǐ

有关建筑结构和设计的物理原理的知识,例如声学。

建筑规程 Jiànzhù guīchéng

［见"建筑法规与标准"］

建房协会 Jiànfáng xiéhuì

［见"储蓄贷款协会"］

建筑标准 Jiànzhù biāozhǔn

［见"建筑法规与标准"］

建筑工种 Jiànzhù gōngzhǒng

建筑业中需要技巧或部分技巧的专业工种。

建成环境 Jiànchéng huánjìng

1. 已有建筑物的区域。
2. 有密集建筑物的城市地区,以区别于自然环境。

建成区 Jiànchéngqū

已普遍开发并有建筑物、仓库及服务性场地的区域,从实体的含义上说,本词现在经常为"城市"的同义词。［参见"聚集体"］

also AGGLOMERATION.

Bulk regulations

In the United States, rules governing the permissible height, depth, density, and width of a building. See also BUILDING CODES AND STANDARDS and FLOOR-AREA RATIO.

Bungalow

In the United States, a small, solidly constructed, one- or one-and-one-half-story house with a veranda. The term is often used to describe any small house.

Buydown

In the United States, a practice for home financing under which the builder or developer of real property absorbs a portion of development costs to provide incentives to buyers. The developer may pay a lending institution a percentage of the interest rate for a property mortgage or make available mortgage money to buyers at reduced rates. Also called a builder discount.

Bypass

In the United States, a road which skirts a built-up area, expediting the movement of traffic and reducing traffic congestion in a town. Generally, the bypass takes the form of a loop joined to a major road. See also EXPRESSWAY.

体积规定　Tǐjīguīdìng

在美国，指控制建筑物的允许高度、深度、密度和宽度所作的规定。[参见"建筑法规与标准"、"建筑面积比"]

带回廊的小住宅　Dàihuílángde xiǎozhùzhái

在美国，指一层或一层半、建造坚实且带回廊的小住宅，有时也可泛指任何小住宅。

折扣售房　Zhékòu shòufáng

在美国，指筹措购房资金的一种方法，由房地产开发者或建造者承担一部分建设费用，以吸引购房者，建造者可在房地产抵押中代买主支付一部分贷款利息或替买主获得低息抵押贷款。又称建造者折扣。

绕行路　Ràoxínglù

在美国，指为了加速交通、减少城市交通拥挤而环绕建成区外围的道路，通常呈半环状，并与主要道路相连。[参见"高速公路"]

C

Cadastral survey

In the United States, an official inventory of the quantity, value, and ownership of real estate. The survey in used for setting individual property tax amounts. See also ASSESSED VALUE.

Capacity building

In the United States, improving the ability of local government to meet the demands of new programs and functions through such means as increasing management efficiency and encouraging community participation in government. See also CITIZEN PARTICIPATION, LOCAL CAPACITY, and USER FEE.

Capital construction

In China, the expanded reproduction of fixed assets in national economic sectors, such as the construction of factories, mines, railroads, bridges, farm irrigation systems, housing, schools, and hospitals, as well as the purchase of machinery and equipment, vehicles, ships, and so forth. See also CAPITAL CONSTRUCTION PROJECT, CONSTRUCTION PROJECT, and INVESTMENT IN CAPITAL CONSTRUCTION.

Capital construction project

In China, a project that has been approved by the

地籍册　Dìjícè

在美国，官方登记有关房地产的数量、价值及其房主的正式记录册，用以核定每块土地应收的地税。[参见"房地产课税价值"]

扩大能力　Kuòdà nénglì

在美国，为适应新的计划和功能的需要而增强地方政府的能力，如通过提高管理效率、鼓励社团参加政府工作等。[参见"市民参与"、"地方能力"、"用户费"]

基本建设　Jīběn jiànshè

在中国，指国民经济各部门固定资产的扩大再生产，如工厂、矿山、铁路、桥梁、农田水利、住宅、学校、医院等工程的建造以及机器设备、车辆、船舶等的购置。[参见"基本建设项目"、"建设项目"、"基本建设投资"]

基本建设项目　Jīběn jiànshè xiàngmù

在中国，指经国家或主管部门批准，用基本建设

government or a ministerial agency and carried out with investment funds for capital construction. See also CAPITAL CONTRUCTION and CONSTRUCTION PROJECT.

Capital equipment

Machinery and devices used in the production process of a company and considered a fixed asset.

Capital expenditures

Amounts of cash or borrowed funds invested to improve or add to land, buildings, machinery, and equipment, or even infrastructure. Capital expenditures are frequently contrasted to operating expenses, i.e., the cost of day-to-day operations.

Capital formation

In accounting, additional investment in new capital assets, e.g., buildings and equipment, calculated after depreciating existing assets. Capital formation is considered essential to economic growth, affecting the course of fluctuations in the business cycle.

Capital grant

In the United States, a cash contribution to support a capital construction project such as water and sewer lines. The capital grant is usually provided by the federal government to a state or local government.

投资进行新建、扩建的工程项目。[参见"基本建设"、"建设项目"]

资本设备 Zīběn shèbèi

企业在生产过程中所需的机器设备,是一种固定资产。

固定资本投资 Gùdìng zīběn tóuzī

为改进土地、建筑物、机器设备或基础设施而投资的现金或贷款,常与经营开支(即日常经营费用)相对而言。

资本形成 Zīběn xíngchéng

在会计学中,指现有资产折旧后进行追加投资以形成新的资产,如建筑物和机器设备。资本形成是经济增长的基础,影响到商业盛衰的循环。

公共工程项目投资 Gōnggòng gōngchéng xiàngmù tóuzī

在美国,指用来资助建设项目(例如给排水管道)的现金补助,常由联邦政府拨给州政府或地方政府。

Capital improvement program

In the United States, a business or government plan enumerating capital expenditures for each year over a period of several years. The plan lists each capital project and identifies beginning and ending dates, anticipated expenditures, and proposed methods for financing expenditures.

Capital investment

Expenditures of an organization or individual to purchase durable assets such as buildings as a means of improving the financial position or furthering the business interests of that individual or organization.

Car pool

In the United States, a transportation arrangement in which a number of persons ride to and from work, school, or recreation together in the same automobile or van, thus reducing fuel costs and highway traffic. See also COMMUTER and PRIORITY LANE.

Carrying capacity

The capability of land and natural environments to sustain uses and growth of certain types. This ecological concept, which makes possible general identification of limits on land use, is critical for land use planning. See also DESIGN WITH NATURE and ENVIRONMENTAL DESIGN.

资本刷新计划 Zīběn shuāxīn jìhuà

在美国,指由企业或政府列出在若干年内每年用于固定资产投资的计划,其中包括全部项目及每项的开始与结束日期、预期费用、有关经费筹措方法的建议等。

资本投资 Zīběn tóuzī

团体或个人为改善其经济地位或增进商业利润而购置耐用资产(如建筑物)的支出。

合乘 Héchéng

在美国,数人约定在上班、上学或娱乐时合乘一辆小轿车或面包车,以便减少汽油费及交通拥挤。[参见"通勤者"、"优先车道"]

持续能力 Chíxù nénglì

土地和自然环境在某些方面经受使用并保持生长的能力。这种生态学上的概念能大致鉴别土地使用的限度,对于土地使用规划十分重要。[参见"适应自然的设计"、"环境设计"]

Carrying charges

In the United States, the expenses incurred from owning idle, interim-use, or nonproductive property, such as the taxes on idle land or on property under construction, the cost of protective services, and insurance.

Cash flow

In general usage, the earnings of a company or individual after interest and tax payments have been deducted. Negative cash flow indicates losses rather than earnings. In the context or real estate investment, cash flow is the disposable cash income from income-producing property that remains after operating expenses and debt service have been deducted from the gross income. Cash flow may also be referred to as income stream.

Catchment area

Originally, a low-lying area which accumulates and holds water from adjacent areas. In planning and sociology, the term refers to areas from which institutions draw their member, and shopping, cultural, social, and recreational centers draw their customers.

Cave dwelling

In China, a dwelling with an arched ceiling formed

土地持有费　Tǔdì chíyǒufèi

在美国，指因产业闲置而需缴纳的费用。例如对闲置不用的空地或正在施工中的房产所征的税收，以及对该产业的保护性服务、保险等所产生的费用。

现金流量　Xiànjīn liúliàng

一般指企业或个人在扣除应付利息及税款后的收益。负现金流表示亏损。在房地产投资中，是指从有收益的产业的总收益中，扣除经营费并清偿债务后所余的可供支配的现金收益。现金流量也可称收入流量。

吸引地区　Xīyǐn dìqū

原意指汇水的低洼地。在城市规划与社会学中，指学术团体的会员或商业、文化、社交与娱乐中心的顾客的集中地。

窑洞住宅　Yáodòng zhùzhái

在中国，指在山区或黄土地区内由水平方向挖洞

by cutting horizontally into a precipice in a loess plateau or hilly region. One type groups a number of dwellings along the walls of a sunken atrium with a common stairway to the ground level above. Cave dwellings of a second type are arranged in tiered rows on steep mountainsides. Variants of such structures have existed for centuries, and similar underground structures are also used for grain storage. See also EARTH-SHELTERED HOUSING.

CBD

See CENTRAL BUSINESS DISTRICT.

CDC

See COMMUNITY DEVELOPMENT CORPORATION.

Cellular growth

Expansion of a city by growth outward as distinct from expansion through separate satellite communities. Cellular growth involves continuous additions on the periphery of the city, whereas expansion through satellites generally implies that open space is left between the city and the new communities.

Census block

In the United States, a designated census area consisting of a city block defined by streets or other physical features. The census block is the smallest

形成的洞形住宅。有在下沉式内庭周围挖洞形成住宅，亦有在陡峭山坡上挖出成排住宅。此类构造已有好几个世纪的历史，除用作住宅外，还可用来储存粮食。[参见"掩土住房"]

商业中心区（缩写） Shāngyè zhōngxīnqū (suōxiě)

[见"商业中心区"]

社区开发公司（缩写） Shèqū kāifā gōngsī (suōxiě)

[见"社区开发公司"]

细胞型增长 Xìbāoxíng zēngzhǎng

城市增长的一种形式，即沿着城市周围向外扩展，犹如细胞的增殖，以区别于卫星城式的增长，即在城市与新社区之间有空地相隔。

人口普查街区 Rénkǒu pǔchá jiēqū

在美国，指人口普查中的一个指定范围。包括以街道或其他实体特征为界的一个街区，是统计分析的最小单位。[参见"人口普查区段"]

geographical unit for statistical analysis. See also CENSUS TRACT.

Census tract

In the United States, a small section of a large city or its adjacent area that has been defined as a unit for statistical purpose and is composed of multiple census blocks. The tract is designed to achieve uniformity of population characteristics, economic status, and living conditions, and to permit comparisons over time. See also CENSUS BLOCK.

Central business district (CBD)

In the United States, the main area of a city or town in which businesses and commercial activities are concentrated. See also DISTRICT.

Central city

1. In China, a city playing the leading role in the politics, economy, or culture of an economic region. See also ECONOMIC REGION.
2. In the United States, the primary city in a Metropolitan Statistical Area, which frequently derives its name from that city. See also METROPOLITAN STATISTICAL AREA.

Central Place Theory

The widely known theory of urban growth originally

人口普查区段 Rénkǒu pǔchá qūduàn

在美国，指大城市或邻近地区的一小部分，划定为统计的单位，由几个人口普查街区组成。区段是根据人口特点、经济情况、生活水平的一致性划分的，便于在较长时期内作比较。[参见"人口普查街区"]

商业中心区 Shāngyè zhōngxīnqū

在美国，指城镇中商业和商务活动集中的主要地区。[参见"区"]

中心城市 Zhōngxīn chéngshì

1. 在中国，指经济区内在政治、经济或文化方面起中心作用的城市。[参见"经济区"]

2. 在美国，指大都市统计区内的主要城市。一个大都市统计区往往以其主要城市的名称命名。[参见"大都市统计区"]

中心地带理论 Zhōngxīndìdài lǐlùn

由 W. 克里斯托勒创始的城市增长理论。所谓"中

New England Colonial Style
Ogden House, Circa 1700
Fairfield Connecticut

美国初期式住宅

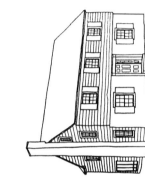

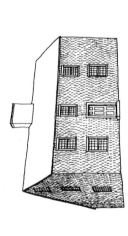

Colonial Revival Style
Hennessy House, 1933
Upper Montclair, New Jersey
(Architect: Arthur Ramhurst)

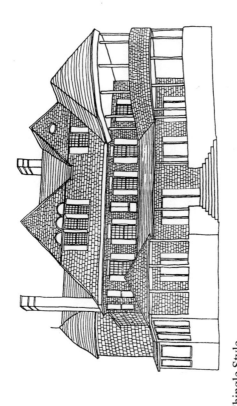

Shingle Style
Isaac Bell Jr. House, 1882
Newport, Rhode Island
(Architects: Mckim, Mead and White)
木瓦式住宅

developed by W. Christaller. The term "central place" refers to the source of goods and services for surrounding areas. The theory implies a complementary relationship between the central and surrounding areas and a specific set of conditions governing the spatial distribution of central places and their hierarchical arrangement. According to the theory, the growth of the city depends on its specialization in urban service functions, while the level of demand over the service area determines the growth rate of the central place. See also LOCATION THEORY.

Centralization

See DECENTRALIZTION.

Centrally administered municipality

See MUNICIPALITY.

Certificate of conformity

A document or mark officially attesting that a product of a service complies with given technical specification. See also CERTIFICATION OF MATERIALS AND PRODUCTS.

Certificate of occupancy

In the United States, a document issued by a government agency or organization to certify that

心地带"系指周围地带的物资及服务的来源地。该理论是关于中心及周围地带的互补关系,以及中心地带空间分布及其分等分级的一系列特定条件。根据这一理论,城市的增长有赖于其城市服务功能的专业化,而其增长速度则取决于其服务地带的需求水平。[参见"位置理论"]

向心化 Xiàngxīnhuà

[见"权力分散"]

直辖市 Zhíxiáshì

[见"市"]

合格证明书 Hégé zhèngmíngshū

证明产品或服务项目与规定的技术规格相符的文件或标记。[参见"材料与制品鉴定"]

符合使用证明书 Fúhé shǐyòng zhèngmíngshū

在美国,由政府机关签发,证明建筑物的全部或部分与有关规范相符,业主或租户可按规定用途

all or a portion of a building meets with applicable building requirements and to authorize use of the building by owners or tenants for its specified purpose. Also called an occupancy permit.

Certification for payment

In the United States, a statement from the architect to the owner confirming the amount of money due to the contractor for work accomplished, or for materials and equipment suitably stored, or both.

Certification of historic buildings

Official confirmation by the U.S. Secretary of the Interior that a property is of historic significance and is protected. The owner receivers tax benefits if the structure is rehabilitated according to standards of the Secretary. However, restrictions on rehabilitation or change cannot be imposed without the owner's agreement, as the federal government does not have zoning power in local jurisdictions. See also HISTORIC AMERICAN BUILDINGS SURVEY, HISTORIC DISRICT, and HISTORIC PRESERVATION.

Certification of materials and products

Favorable technical evaluation of the fitness for purpose of materials, assemblies, and workmanship. Technical assessment of performance may take

使用的文件。亦称使用许可证。

付款证明书 Fùkuǎn zhèngmíngshū

在美国，指建筑师致业主的文件，证明因已完成的工程与（或）已妥善存储的材料、设备而应付予承包人的款项。

历史性建筑保护证明 Lìshǐxìng jiànzhù bǎohùzhèngmíng

美国内政部长对具有历史价值应予保护的建筑物所给予的正式确认。如该建筑物根据内政部所订标准加以修建，则业主可得到减税优待。但因联邦政府在地方法律上并无区划权，故如不征得业主同意，无权强行修建或改建。[参见"美国历史性建筑调查"、"历史性市区"、"历史性建筑"]

材料与制品鉴定 Cáiliào yǔ zhìpǐn jiàndìng

对材料、组装和加工精度所做的技术评定。对功能的技术性鉴定可包括安全、适用、耐久和预期用途等因素。凡符合特定技术规定的产品即被认

into account such factors as safety, durability, and intended purpose. A product that is judged to be in compliance with a given technical specification is considered certified. See also ACCREDITATION, BUILDING MATERIAL TESTING, CERTIFICATE OF CONFORMITY, QUALITY CONTROL, and TECHNOLOGY ASSESSMENT.

Change in the work

See CHANGE ORDER.

Change order

In the United States, written instructions to the contractor signed by the owner and the architect of a building constructed under contract. The order authorizes changes in the work, e.g., additions, deletions, or revisions, with corresponding adjustments in the amount paid the contractor or the term of the contract. Modifications to the contract sum, e.g., deductions, or to the contract time are only possible through change orders. See also CONSTRUCTION COST and CONTRACT.

Citizen participation

In the United States and Great Britain, involvement of the population of a neighborhood, town, or city in activities or projects beneficial to the whole group, such as planning and implementation of community

可。[参见"确认"、"建筑材料检验"、"合格证明书"、"质量控制"、"技术性鉴定"]

工程变更 Gōngchéng biàngēng
[见"工程变更通知单"]

工程变更通知单 Gōngchéng biàngēng tōngzhīdān
在美国,指在根据合同建造的工程中,由业主和建筑师签发给承包人的书面通知,授权承包人对工程作某些更改,因而对应付工程款项或合同条款作相应的调整。对合同总价及施工期限的修改只能通过工程变更通知单方能进行。[参见"建筑成本"、"合同"]

市民参与 Shìmíncānyù
在美国和英国,邻里、市镇或城市的居民参与有关公益的活动或计划,例如对社区发展计划的规划与实施等。亦称"社区参与"、"邻里参与"、"公众参与"、"社会参与"。[参见"扩大能力"]

development programs. Also called community participation, neighborhood participation. See also CAPACITY BUILDING.

City Beautiful Movement

An architectural movement sparked by the Chicago World's Columbian Exposition of 1893. The purpose was to bring order to the confusion characteristic of American urban areas by redesigning cities to create a sense of order and harmony. Boulevards and grand-scale buildings were constructed in keeping with the beaux-arts aesthetic of the times. The movement popularized city planning and resulted in the replanning of a number of American cities.

City district

See DISTRICT.

City district planning

In China, preparation of a plan for a city district after comprehensive planning has been completed. Each district in the city has a different plan, depending on the district's nature and function. See also DISTRICT and URBAN PLANNING.

City manager

In the United States, an official appointed by an elected council to administer a city government

城市美化运动　Chéngshì měihuà yùndòng

由1893年在芝加哥举行的世界哥伦比亚博览会所引起的一种建筑运动，旨在整顿美国城市地区的混乱情况，通过重新设计创造和谐秩序。据此，按当时学院派的美学观点建造了林荫大道和大尺度的建筑物。这次运动推动了城市规划，使美国若干城市重新修订规划。

市区　Shìqū

［见"区"］

城市分区规划　Chéngshì fēnqū guīhuà

在中国，指根据已编制的城市总体规划所做的市内各局部地区的规划。按不同的功能和性质，各区有各自的规划图。［参见"区"、"城市规划"］

市行政官　Shì xíngzhèngguān

在美国，根据政府的"议会—行政官"制度，由市议会任命的管理市政府的官员。在县一级的相

under the council-manager plan of government. The equivalent official at the county level is called a county manager.

City planning and administration bureau

In China, a government agency below the mayoral level which is responsible for making up, implementing, and administering the urban plan.

City planning commission

An agency within city government with urban planning and land use planning functions. In China, the commission is a high-level agency that makes policy decisions for planning, sometimes also supervising preparation and implementation of plans for land use and physical development. In the United States, the agency, which is often in the office of the mayor, is responsible primarily for preparation of comprehensive plans. Recommendations of the commission generally have to be approved by the board of zoning appeals. See also BOARD OF ZONING APPEALS and PLANNING COMMISSION.

City proper

In China, the central part of the city, not including suburbs, as defined by the city government.

应官员称县行政官。

城市规划管理局　Chéngshì guīhuà guǎnlǐjú

指中国主管城市规划的编制、实施与管理的部门，属市政府领导。

城市规划委员会　Chéngshì guīhuà wěiyuánhuì

市政府下设管理城市规划和土地使用规划的机构。在中国，委员会为对规划工作进行高级决策的机构，有时也对土地利用及具体建设规划的准备和实施进行监督。在美国，通常为市长办公室的下设机构，主要负责总体规划的编制。委员会所提出的建议，一般需经申诉委员会批准。[参见"区划申诉委员会"、"规划委员会"]

城区　Chéngqū

在中国，由市政府划定的城市中心与周围地区，不包括郊区。

City size

The size of a city is usually measured by the number of inhabitants, although the land area that it covers may be taken as a secondary measure. In China, a metropolis, usually the main city and/or capital of a province or region, has a population of over 1,000,000 inhabitants; a large city, 500,000 to 1,000,000; a medium-sized city, 200,000 to 500,000; and a small city, under 200,000. For U.S. census purpose, urban settlements are classified as cities if the population exceeds 2,500 inhabitants and as places if the population numbers less than 2,500.

Cityscape

The general appearance and atmosphere created by the streets, squares, buildings, and landscapes of a city. See also SKYLINE, STREETSCAPE, and URBAN DESIGN.

Civic building

A structure in the United States used for government administrative offices, libraries, courts, galleries, and other government-provided public services. See also CIVIC CENTER.

Civic center

In the United States, that part of a town or city, usually located centrally, with buildings for civic

城市规模 Chéngshì guīmó

通常指城市的人口数量。有时以城市用地面积为辅助标志。在中国，人口超过100万为特大城市（常为省或自治区的重要城市与/或首府），人口在50万至100万的为大城市；人口在20万至50万的为中等城市；人口在20万以下的为小城市。在美国，人口超过2500人称为城市，低于2500人称为地方。

城市风貌 Chéngshì fēngmào

城市中由街道、广场、建筑物、园林绿化等形成的体形外观及气氛。[参见"天际线"、"街景"、"城市设计"]

市政建筑 Shìzhèng jiànzhù

在美国，指政府机构的办公楼、图书馆、法院、美术馆及其他政府提供的为公众服务的建筑物。[参见"市政中心"]

市政中心 Shìzhèng zhōngxīn

在美国，指城市中行政建筑（如市政大厅、行政办公楼）、法院及文化娱乐场所等集中的城市中心

administration. See also CIVIC BUILDING.

Civil building

In China, any structure designed for living or social activities rather than for industrial production, classified either as a residential or public building. See also INDUSTRIAL BUILDING and RESIDENTIAL BUILDING.

Civil defense planning

In China, a constituent part of comprehensive planning, encompassing distribution of necessary underground facilities with exits, entrances, and passages that are coordinated with above-ground structures and under-ground piping systems. In peacetime, such structures are used as cafeterias and as cool air reservoirs to be used as a natural cooling system for buildings, especially for movie theaters. In the United States, civil defense planning is coordinated at the federal level as part of overall emergency planning, but carried out by local governments.

Client

In the Western business sector, the individual or organization to whom/which and individual contracting organization supplies a service or product. See also OWNER.

地段。[参见"市政建筑"]

民用建筑 Mínyòng jiànzhù

在中国,指用于居住或社会活动的建筑物,不包括工业生产用房。分为居住建筑和公共建筑。[参见"工业建筑"、"居住建筑"]

城市人防规划 Chéngshì rénfáng guīhuà

在中国,指城市总体规划的一个组成部分,内容包括:必不可少的地下建筑;防空通道及安全出入口等的布置;与地面以上建筑物及地下管道网的协调;对于这些构筑物的平时利用,例如用作餐室及冷风源;紧急状态下的疏散和急救措施等。在美国,人防规划是全面应急规划的组成部分,在联邦一级取得协调,但由地方政府实施。

客户 Kèhù

西方商业中承包者或承包机构提供服务或产品的对象。[参见"业主"]

Closed system

Parts or materials that have not been manufactured according to specific standards so that substitution of equivalents in the process of building is not possible. See also DIMENSIONAL COORDINATION and OPEN SYSTEM.

Closing

The conclusion of a transaction. In real estate, closing includes the delivery of a deed, financial adjustments, the signing of notes, and the disbursement of funds necessary to the sale or loan transaction.

Closing costs

See SETTLEMENT COSTS.

Cluster development

A housing development pattern in the United States in which structures are placed in closely related groups around cul-de-sacs, courts, and loop streets. The houses are arranged to take advantage of the natural features of the land, to provide for green spaces, variety of structures, and a sense of community. Economic objectives are to reduce land, energy, and public service costs. Normally, high residential density is permitted, and a large proportion of the land is used for common open space, which is managed by the community

专用建筑体系 Zhuānyòng jiànzhù tǐxì

不按专业标准制造的建筑部件或材料,因而在施工中不得以相应部件或材料替代。[参见"尺寸协调"、"通用建筑体系"]

结算 Jiésuàn

交易的终结。在房地产交易中,结算包括交付契据、财务整理、票据签字,并支付销售或货款交易所需资金。

结算费用 Jiésuàn fèiyòng

[见"成交费用"]

组团式建设 Zǔtuánshì jiànshè

美国的一种住宅建设方式,建筑围绕尽端路、庭院和环形路成群布置,并利用自然地形安排绿地。建筑物的形体有变化,具有居住社区的气氛。这种建设方式的经济目标是减少用地、能源及公共服务费用。在一般情况下准许较高的居住密度,大部分空地都用作公共绿地,由社区协会管理。[参见"组团式区划"、"社区协会"、"有规划的地段建设"]

association. See also CLUSTER ZONING, COMMUNITY ASSOCIATION, and PLANNED UNIT DEVELOPMENT.

Cluster zoning

In the United States, an approach to zoning that allows developers to concentrate housing units in one portion of a land tract and to leave the remainder of the land as open space or park land. Cluster development is designed to maximize open space in residential developments. See also CLUSTER DEVELOPMENT and DENSITY TRANSFER.

CM

See CONSTRUCTION MANAGEMENT.

Code enforcement

In the United States, steps such as inspection, licensing, and issuance of permits to ensure observance of regulations established by an official authority for a particular activity such as construction. See also BUILDING CODES AND STANDARDS, BUILDING INSPECTOR, UNAUTHORIZED CONSTRUCTION, and ZONING ADMINISTRATOR.

Coefficient of utilization

For public utilities, the ratio of the peak load to the

组团式区划 Zǔtuánshì qūhuà

美国的区划方法之一,允许建设者在一片土地上集中建造住宅,而将所余部分用作绿地或公园,其目的在于在居住区内取得最大限度的空地。[参见"组团式建设"、"密度调剂"]

施工管理(缩写) Shīgōng guǎnlǐ (suōxiě)

见["施工管理"]

执行法规 Zhíxíng fǎguī

在美国,指为了保证官方所制定的法规能得到遵守而采取的步骤,例如检查、注册及颁发执照等。[参见"建筑法规与标准"、"建筑监察员"、"违章建筑"、"区划执行官"]

利用系数 Lìyòng xìshù

在公共设施中,一个工厂或一种系统在高峰时的

installed capacity of a plant or system.

Cofinancing

Loan money for a project or purchase obtained from more than one lender.

COG

See COUNCIL OF GOVERNMENTS.

Coinsurance

In the United States, a method of insuring property which requires that the insured party carry insurance equal to a percentage of the insured property's value or share the loss if a claim is filed. See also PROPERTY INSURANCE.

Collateral

In the United States, personal or real property or securities provided by a borrower to the lender to guarantee repayment of a loan.

Collector street

In the United States, a principal roadway within a residential area that conducts traffic between major arterial streets, carries a relatively high volume of vehicles, and supports minor commercial establishments along its route.

负荷与安装容量之比。

共筹资金 Gòngchóu zījīn

为进行一项工程或购置一项产业而由几个贷款人同时提供贷款。

规划协调委员会（缩写） Guīhuà xiétiáo wěiyuánhuì (suōxiě)

［见"规划协调委员会"］

共同保险 Gòngtóng bǎoxiǎn

美国财产保险的一种形式，规定被保一方承担一定百分比的财产价值保险，或在申请赔偿时分担一部分损失。［参见"财产保险"］

担保品 Dānbǎopǐn

在美国，指由借款人向贷款人提供的动产、不动产或证券，作为偿还借款的抵押品。

辅助道路 Fǔzhù dàolù

指美国居住区内的主要街道，用以疏导主干道之间的交通，承担着相当大的交通量，并有利于沿路小型商业设施的发展。

Commercial district

See DISTRICT.

Commercial network

In China, different levels of commercial organizations (e.g. district, subdistrict) that are located around the city at strategic points to maximize efficiency and productivity. Each strategic point has its own defined function in the system of commercial establishments. Designating the distribution, location, function, scope, and interrelationships of commercial and service organizations such as grocers and greengrocers, retail shops, drug stores, post offices, and savings banks, is part of comprehensive planning.

Commercial space

Areas within a building designated for rental to businesses.

Commitment

1. An agreement, often in writing, between a lender and a borrower for the loan of money at a future date, subject to compliance with stated conditions.

2. An agreement by a lender to provide long-term financing to a builder secured by an existing or proposed building. The commitment usually provides for later substitution of a to-be-approved owner/occupant at a higher loan amount than committed to

商业区 Shāngyè qū

［见"区"］

商业网点 Shāngyè wǎngdiǎn

指中国城市中商业与服务业机构的网络。一般按分等分级（例如市区和小区级）设置的原则，即在全市或一定地域范围内布点，以发挥最大的经济效益，并确立每个点的作用与经营范围，以及点与点之间的分工合作关系。商业网点的规划是总体规划的一部分，其经营服务内容有副食蔬菜店、零售商店、药店、邮局、储蓄所等。

商业面积 Shāngyè miànjī

建筑物内可出租供营业使用的面积。

承诺 Chéngnuò

1. 指在借方与贷方之间，按陈述的条件在未来某一日期借款的书面协议。

2. 指贷方同意给营造商以长期资助的协议，以一栋建成的或拟建的房屋作为担保。协议通常也规定，允许日后愿提供更高的贷款者拥有或使用该房屋。

the builder.

Common-area maintenance

In the United States, upkeep of areas shared by residents of condominiums, housing cooperatives, or townhouses. A fee is usually charged for this service. See also CONDOMINIUN, HOUSING COOPERATIVE, PLANNED UNIT DEVELOPMENT, and TOWN-HOUSE.

Community

A term applied in American English to a group of people living together as a social unit and sharing common interests and goals related to activities and facilities that serve the public. "Community" implies unity for certain purposes and willingness to provide mutual support for those purposes, formally through taxation and informally through volunteer efforts of group members. The term is usually used to describe towns or small cities as social rather than administrative units, but may also be applied to larger units, including the "community of nations."

Community association

In the United States, a nonprofit organization composed of homeowners or renters that often manages common open spaces and recreation facilities under contract in a planned unit development or

公用部分的维护 Gōngyòngbùfènde wéihù

在美国,指住户自有公寓、合作住房或市镇联立式住宅中对住户共用的地方所做的维护。通常收取维护费。[参见"住户自有公寓"、"合作住房"、"有规划的地段建设"、"市镇联立式住宅"]

社团、社区 Shètuán、shèqū

在美国英语中,该词指住在一地,在服务性的活动和设施上有着共同利害关系的一组社会单元。并在某些目的和意愿上,通过正式的纳税手段,或通过非正式的自愿努力,达到团结互助。该词有时也用来指社会概念而非行政概念的小城镇,甚至可以扩大指国家间的共同体。

社区协会 Shèqū xiéhuì

在美国,指由房主或租户组成的非营利性组织,常以承包方式管理有规划的地段建设或新城中的公用绿地和娱乐设施。[参见"美观控制"、"组团式建设"、"有规划的地段建设"]

new town. See also AESTHETIC CONTROLS, CLUSTER DEVELOPMENT, and PLANNED UNIT DEVELOPMENT.

Community center

In the United States, a facility within a town, residential area, or apartment development designed as a place for residents to meet and to participate together in recreational, educational, and social activities. See also COMMUNITY FACILITY.

Community college

In the United States, an educational institution run on a local level to afford the general public ready access to instructive courses in a variety of fields and to provide recent high school graduates with a high-quality, low-priced alternative to the first two years of college or university studies or preparation for a specific vocation.

Community development

1. The growth processes of a city or town.
2. In the United States, the planned use of urban land for improvement or expansion.

Community development corporation (CDC)

In the United States, an organization formed by

社区中心　Shèqū zhōngxīn

在美国，指在村镇、居住区或公寓建筑群中的一种设施，作为居民集会、娱乐、教育和社会活动的场所。[参见"社区设施"]

社区大学　Shèqū dàxué

在美国，指由地方主办的教育机构，为公众就近提供内容广泛的课程，并为高中毕业生提供质量高而收费低廉的相当于大学一、二年级程度的课程或专业培训。

社区建设　Shèqū jiànshè

1. 城市或市镇的发展过程。
2. 在美国，指为改进或扩展城市而有计划地使用城市土地。

社区开发公司　Shèqū kāifā gōngsī

在美国，由不景气地区的居民组织的团体，旨在

residents of impoverished or depressed areas to stimulate, finance, and even operate local businesses, and to create jobs at a local level.

Community facility

In the United States, a structure within a city or town designed to meet community needs for particular types of services, for example, a school, hospital, restaurant, or retail store. See also COMMUNITY CENTER, HALFWAY HOUSE, and INFRASTRUCTURE. For the Chinese equivalent, see also SERVICE FACILITY.

Community participation

See CITIZEN PARTICIPATION.

Commuter

A person who regularly travels some distance by automobile, bus, train, or airplane, usually from residence to place of employment and back. Commuting is most commonly an urban phenomenon, linking the residential suburbs to city business centers, but may occur in a variety of other situations, including city to city. See also CAR POOL.

Compensation for land

In the United States, acquisition of land by the

促进、筹措或经营该地区的商业并增加就业机会。

社区设施 Shèqū shèshī

在美国,城市或城镇中为适应居民需要而设的各种服务设施,如学校、医院、餐馆、零售店等。[参见"社区中心"、"戒瘾教育所"、"基础设施",中国相应的词参见"服务设施"]

社区参与 Shèqū cānyù

[见"市民参与"]

通勤者 Tōngqínzhě

定期在住所与工作地点之间乘坐汽车、公共汽车、火车或飞机远途往返者。通常为一种城市现象,即往返于郊区的住所和市区商业中心的工作地点之间。但也有各种不同情况,如往返于两座城市之间。[参见"合乘"]

土地补偿费 Tǔdì bǔchángfèi

在美国,根据国家征地权而由政府征用土地时,

government under the right of eminent domain requires payment to the landowner for the fair market price of property. See also EMINENT DOMAIN and LAND ACQUISITION.

Compensatory replacement of demolished housing

Policy in China according to which a private house demolished by a development organization for a development project must be replaced with a structure of comparable size and shape on another suitable site. See also LAND ACQUISITION and RELOCATION HOUSEHOLD.

Competitive funding

Allocation of grant money to states by the U.S. federal government as a means of promoting new initiatives to understand and resolve social problems of national scope. No equal distribution requirements are set, and grantees must compete for funds. See also BLOCK GRANT, DISCRETIONARY FUNDING, and FORMULA FUNDING.

Completed item

In China, one of the main statistical indices of capital construction. The term refers to a production (or major) item of a construction project, wholly or partially completed, that has fully or partially realized

按公平的市场价格付给业主的补偿费。[参见"国家征地权"、"征地"]

原拆原建 Yuánchāi yuánjiàn

在中国,指因城乡建设需要而拆除私人住房及院落,由征地单位另选适宜地址按其原状重建补偿。[参见"征地"、"拆迁户"]

竞争性拨款 Jìngzhēngxìng bōkuǎn

由美国联邦政府批给各州的拨款,旨在提高各州对于了解及解决全国性的社会问题的积极性。这种拨款并无等量分配的规定,而需由各州来争取。[参见"一揽子赠款"、"选择性拨款"、"按公式拨款"]

建成投入生产项目 Jiànchéng tóurù shēngchǎn xiàngmù

在中国,是基本建设主要统计指标之一。建设项目中的生产性(或主要)工程项目全部或部分建成,达到设计规定的全部或部分生产能力(或效益),经验收移交生产(或使用)的工程项目。

the production capacity or performance efficiency for which it was designed and that has been accepted and handed over for assignment or occupancy. See also ACCEPTANCE OF WORK SUBELEMENTS.

Completed project

In China, one of the main statistical indices of capital construction. The term refers to the sum total of production and nonproduction items of a construction project, which have been completed, accepted, and handed over for assignment or occupancy. Completion of the entire project marks the end of the construction process. See also FINAL ACCEPTANCE.

Completion acceptance

See FINAL ACCEPTANCE.

Completion bond

In the United States, an official written guarantee to the lender by a contractor that a project will be completed free of liens. See also BOND and PERFORMANCE BOND.

Complex zone

See MIXED-USE ZONE.

［参见"分项工程验收"］

全部竣工项目 Quánbù jùngōng xiàngmù

在中国,是基本建设主要统计指标之一。整个建设项目中的生产性和非生产性工程项目已全部建成并经验收移交生产(或使用)的建成项目。建设项目的全部竣工,标志着建设全过程的终结。
［参见"最终验收"］

竣工验收 Jùngōng yànshōu

［见"最终验收"］

完工保证书 Wángōng bǎozhèngshū

在美国,指承包人致贷款人的书面保证书,保证工程将按时完工,不遗留未了事项。［参见"保证书"、"履行合同保单"］

综合区 Zōnghéqū

［见"多用途区"］

Comprehensive economic equilibrium

Overall balance in the distribution of labor, materials, and funds among different sectors of the national economy in order to achieve balance between production goals and social demand. The concept is fundamental to the process of planning the national economy in China.

Comprehensive planning

1. In the United States, a formal approach to urban planning which, as a guide for governmental policies and action, includes preparation of general plans on land use, provision of public facilities and services, human and natural resources utilization, and environmental needs; identification of area needs; surveys of historic structures; development of action plans and programming of capital improvements and other expenditures; coordination of all plans and activities; and recommendation of regulatory measures. Zoning is based on both comprehensive planning and on local government policies and is implemented at the city or county level. See also LAND USE SURVEY, MASTER PLAN, URBAN PLANNING, and ZONING.

2. In China, urban planning at the comprehensive level. Comprehensive planning includes designating the function of a city, its size limit and development

综合平衡 Zōnghé pínghéng

指投入国民经济各个部门的人力、物力、财力的全面平衡，其目的在于建立社会生产和生活需要之间的平衡，是中国国民经济计划工作的基本方法。

总体规划 Zǒngtǐ guīhuà

1. 在美国，指城市规划的一种正式做法，作为政府政策和行动的一种指导。内容包括土地使用、公共服务设施的设置、人力和自然资源的利用、环境要求等的全面规划以及地区内各种要求的确定；历史建筑的调查；实施计划的开展与资金和其他开支的计划；各项规划和行动的协调；为实现规划和行动所提出的协调措施。根据总体规划和地方政府的政策制定用地区划，由市或县一级执行。[参见"土地利用调查"、"规划总图"、"城市规划"、"区划"]

2. 在中国，指综合性的城市规划。是确定一个城市的性质、规模、发展方向以及制订城市中各类建设的总体布局和全面环境安排的城市规划。总

orientation, as well as the general disposition of all lines of development and overall consideration of the environment for development. Planning at this level also includes determination of relevant norms and indices, as well as programming of long- and short-term goals with procedures and measures for implementation. See also DETAILED PLANNING and URBAN PLANNING.

Concentric Zone Theory

A theory elaborated by E.W. Burgess which holds that a city expands radially from the center, forming a series of differentiated concentric zones. As growth occurs, each inner zone, with the central business district at the core, extends its area by invading the next outer zone, changing the character of that area. See also LOCATION THEORY.

Concept development

The inception and elaboration of an idea or approach, usually with input from a variety of sources. See also BRAINSTORMING.

Concept plan

A rough sketch or design showing in broad outline the ideas and objectives to be defined in detail in a final plan.

体规划还包括选定规划定额指标、制订该市远、近期目标及其实施步骤和措施等工作。[参见"详细规划"、"城市规划"]

同心区理论 Tóngxīnqū lǐlùn

由 E.W. 伯格斯提倡的理论,认为城市是由中心向四周作辐射状的发展,形成一系列的同心区域。在发展过程中,每一个区域都以中央的商业区为核心,向外围扩展,从而改变外围邻区的性质。[参见"位置理论"]

概念发展 Gàiniàn fāzhǎn

在吸收了各方面的输入后所形成并发展的一种想法或做法。[参见"绞脑汁会"]

概要性规划图 Gàiyàoxìng guīhuàtú

一种粗略的草图或设计,用以表达想法与目的的规划要点,其细节需要在最后的规划图中加以明确。

Condemnation

1. In the United States, declaration by a public authority with the power of eminent domain that a property is needed for public use and will be acquired from the owner for a fair market price, regardless of whether the owner is willing to sell. See also EMINENT DOMAIN.

2. Official declaration by government authorities that a structure is unfit or unsafe for human habitation and should be demolished.

Condominium

In the United States, an estate in real property consisting of an individual ownership in a residential, commercial, or industrial unit, and an undivided but proportional interest in common areas. The term most frequently refers to an individually owned living unit, usually in a multiunit building. Areas such as open spaces, recreation facilities, and hallways are commonly owned and can be used by all residents, who must usually pay a monthly fee for maintenance. See also COMMON-AREA MAINTENANCE and HOUSING COOPERATIVE.

Congregate housing

In the United States, dwellings for the elderly or handicapped that include meals, housekeeping,

1. 征用 Zhēngyòng

在美国,由具有国家征地权的权力机构宣布将某一产业用于公共用途,并不论业主愿意出售与否,以公平的市场价格向业主购取。[参见"国家征地权"]

2. 废弃 Fèiqì

由政府机构宣布某一建筑物不安全或不适宜居住而应予拆除。

住户自有公寓 Zhùhù zìyǒu gōngyù

在美国,指在一栋居住、商业或工业建筑内一套私有的房产。公用部分则按比例共有。该词多指居住单元为私人所有的多单元住宅楼。绿地、娱乐设施、楼厅等属公共所有,住户均可使用,每月需付维护费。[参见"公用部分的维护"、"住房合作社"]

集合公寓 Jíhé gōngyù

在美国,供老年人或残疾者居住的住房。提供膳食、家政服务及对个人的照料,使部分残疾者或

and personal care assistance, enabling partially impaired and/or socially isolated citizens to maintain a semi-independent lifestyle. See also ELDERLY HOUSING.

Conservation

Planned use of wildlife, natural resources, and manmade resources to assure their preservation or replenishment. This includes planning and design of proposed construction to preserve significant existing natural values. See also ENERGY CONSERVATION, RESOURCE MANAGEMENT, and TERRITORIAL PLANNING.

Conservation of historic buildings

In China, the protection of ancient palaces, temples, vernacular dwellings, and other types of structures that have historic value. Such buildings may be situated in Historic Cultural Cities, such as the Imperial Palace in Beijing, but some are not, such as the Zhaozhou Bridge in Hepei province and vernacular dwellings in various localities. Directive orders have been issued by the State Council concerning the conservation of historic buildings. For the United States equivalent, see also HISTORIC PRESERVATION.

在社会上孤独无依的居民得以维持半独立的生活方式。[参见"老年住房"]

保护 Bǎohù

对于野生动、植物以及自然资源、人造资源的利用所采取的一种措施,以确保这些资源得到保护和补充,包括在规划和设计工程时,保护现存具有重要价值的自然资源。[参见"能源保护"、"资源管理"、"国土规划"]

古建筑保护 Gǔjiànzhù bǎohù

中国对于历史上有价值的古代宫殿、寺庙、民居、工程构筑物等进行保护,其中一部分位于历史文化名城范围内,如北京的故宫;一部分则不在其中,如河北省的赵州桥以及各地的民居等。国务院对古建筑保护有专门的指令性文件。[美国的相应词汇,参见"历史性建筑保护"]

Conservation of Historic Cultural Cities

In China, protection of cities that have historic, geological, cultural, or archeological value. Directive orders have been issued by the State Council designating a number of cities as Historic Cultural Cities. In such a city, all aspects of comprehensive planning, including land use, location of industry, and architectural style, must focus on preservation of the city's cultural heritage. As of 1982, the State Council had conferred that status on twenty-four cities, including Beijing, Xi'an, Yangzhou, and Suzhou.

Conservation of historic landmarks and sites

In China, protection of excavated and unexcavated sites and landmarks that are of historic value, including graves such as the Ming Tombs in Beijing and historic sites, stone carvings, and structures such as the Mogao Grottoes in Dunhuang in Gansu Province. Under directive orders issued by the State Council, such historic treasures are major historic landmarks to be protected by the State, province, municipality, or county.

Conservation of scenic spots

In China, the protection of natural scenery and features, including famous mountains, rivers, springs, waterfalls, stone forests and caves, or places that are

历史文化名城保护 Lìshǐ wénhuà míngchéng bǎohù

中国对具有历史、地理、文化和考古等价值的城市进行保护。国务院批准确定一些城市为历史文化名城,并有指令性文件,一经确定为历史文化名城后,该城市的总体规划,包括土地利用、工业布点、建筑风格等,都必须以保护其传统文化为主要目标。1982年经国务院批准,确定北京、西安、扬州、苏州等24个城市为历史文化名城。

文物古迹保护 Wénwùgǔjì bǎohù

中国对于历史上有价值的文物古迹,无论出土与否,都进行保护,包括墓葬、古代遗址、石刻、工程设施等,如北京的明十三陵、甘肃省敦煌的莫高窟。国务院对文物古迹的保护按国家、省、市、县分级,分别确定重点文物保护单位,并有指令性文件。

风景名胜保护 Fēngjǐng míngshèng bǎohù

中国对于优美的自然风景区,包括名山、大川、名泉、瀑布、石林、溶洞等,以及历史上流传至今的名胜进行保护。国务院有加强风景名胜保护

historically famous. Guidelines for their protection and administration are outlined in directive orders of the State Council. Regulations by local authorities restrict development in the vicinity of such spots, including road construction and the height, shape, and style of buildings.

Construction

All the on-site work performed in building or altering structures, from land clearance to completion, including excavation of the site, erection of the structure, and assembly and installation of components and equipment.

Construction and installation work

In China, construction work which includes new construction; building remodeling and repair; installation of heating systems, electrical wiring, plumbing, and gas lines; and construction of foundations for machinery. Installation work consists mainly of installing machinery and equipment for manufacturing, power generation, and heavy lifting.

Construction brigade

The basic administrative unit of a construction enterprise. The brigade has its own separate budget and may report either directly to the construction

和加强管理的指令性文件,各城市还对于在风景名胜区附近的建设如道路建设、建筑高度、建筑造型与风格等作限制性规定。

施工 Shīgōng

在现场进行的全部新建或改建工程,从清理场地到完工,包括挖土、建造以及构件与设备的装配和安装。

建筑安装工程 Jiànzhù ānzhuāng gōngchéng

中国对建筑工程和安装工程的总称。建筑工程主要有:建筑物和构筑物的新建、改建和修复;供暖、卫生、通风、照明、煤气管道敷设等。安装工程主要有:生产、动力、起重等各种机械设备的安装。

工程队(工段) Gōngchéngduì (gōngduàn)

在中国的建筑企业中,承担具体的工程任务,实行内部经济核算的基层管理机构,受建筑公司或工程处(工区)领导。[参见"建筑公司"、"工

company or to the construction division. See also CONSTRUCTION COMPANY and CONSTRUCTION DIVISION.

Construction budget

In the United States, the sum established by the owner as available for construction of a project. The budget is usually equal to the stipulated highest acceptable bid, or in the case of a project involving multiple construction contracts, the stipulated aggregate total of the highest acceptable bids. See also BID and PROJECT BUDGET.

Construction company

1. In the United States, a private business organization that specializes in some type of construction activities, whether for roads, buildings, dams, or bridges.

2. In China, an organization in a construction enterprise, with the status of a juristic person and a separate budget and accounting system. In a three-level management system, the levels in a construction enterprise consist of the construction company, the construction division, and the construction brigade, but only of the construction company and the construction brigade when a two-level management system is adopted. See also CONSTRUCTION

程处"]

工程预算 Gōngchéng yùsuàn

在美国,业主准备用于一项工程上的投资额,预算一般等于规定的最高中选标价。如一项工程涉及多项合同,则为各项中选标价之总和。[参见"投标"、"建设项目投资预算"]

建筑公司 Jiànzhù gōngsī

1. 在美国,指专营施工业务的私营商业性机构,承担建造房屋、道桥或水坝等工程。

2. 在中国,指建筑企业的一级组织机构,具有法人性质,实行独立的经济核算。建筑公司有的实行三级管理,即由公司、工程处(工区)和工程队(工段)组成;也有的实行两级管理,即由公司和工程队组成。[参见"工程队"、"工程处"、"建筑安装企业"]

BRIGADE, CONSTRUCTION DIVISION, and CONSTRUCTION ENTERPRISE.

Construction cost

1. In China, the cost of construction work as determined in a detailed estimate, including direct, indirect, and extra expenses, but excluding the statutory profit paid to a construction organization as part of the construction payment. See also CONSTRUCTION PAYMENT, DIRECT EXPENSES, EXTRA EXPENSES, and INDIRECT EXPENSES.

2. In the United States, the outlay for construction work as determined through the competitive bidding process. Expenses for labor and materials, overhead, and profit are usually calculated in the bid proposal of each contractor. The bid accepted by the owner is essentially the construction cost, although contingency provisions in the contract between the owner and the contractor selected allow for cost adjustments. See also ACCEPPTED BID, BASE BID, CHANGE ORDER, and DETAILED ESTIMATE.

Construction crew

In the United States, a term used loosely for all the workers on a construction site or for the specialized workers who perform a particular type of work

建筑成本 Jiànzhù chéngběn

1. 在中国,根据设计预算价格计算确定的建筑成本,包括直接费、间接费、独立费,但不包括支付给施工单位的法定利润。[参见"建筑造价"、"直接费"、"独立费"、"间接费"]

2. 在美国,指通过竞争性投标确定的施工费用,包括人工、材料、管理费及利润,由承包商开列在投标书中。中标价基本上即为施工费用。但合同允许对意外情况在费用上作一定调整。[参见"中选标"、"基本标价"、"工程变更通知单"、"设计预算"]

施工队伍 Shīgōng duìwǔ

该词在美国并无严格定义,可泛指在建筑工地上的全部工人,亦可指某一工程的专业工人。[中国的相应名词,参见"施工队"]

on the site. For the Chinese equivalent, see also CONSTRUCTION TEAM.

Construction division

In China, the administrative subdivision of a construction organization, generally maintaining a separate budget and reporting to the construction company. See also CONSTRUCTION BRIGADE and CONSTRUCTION COMPANY.

Construction document design

1. In China, the final phase of architectural design work based on the approved results of preliminary design or design development. Construction document design involves completion of extensive working drawings, showing the dimensions, materials, construction details, and installation work, as well as specifications and estimates.

2. In the United States, the final stage of the design phase. Construction documents drawn up from the approved design development documents contain the level of precise detail required for construction, including cross-section drawings, definitive material specifications, physical layout of all systems, and instructions for installation of fixtures such as windows. As an important support activity for the process of contracting out construction work,

工程处（工区） Gōngchéngchù (gōngqū)

在中国的建筑企业内，受建筑公司领导，实行内部经济核算的一级中层管理机构。[参见"工程队"、"建筑公司"]

施工图设计 Shīgōngtú shèjì

1. 在中国，指建筑设计工作的最后阶段。根据批准的初步设计（或技术设计）编制，内容包括确定全部工程尺寸、用料、结构、构造、设备等的施工图纸、说明书和预算。

2. 在美国，指设计过程的最后阶段。在技术设计文件基础上制定的施工图设计文件，提供施工要求的准确细节，包括断面图、确定的材料规格、各个系统的布置图和固定装置如窗户的安装说明。作为发包施工工程过程中的一项重要辅助活动，施工图预算常将施工工程的各组成部分分类汇总，例如平土和开挖工程或基础和土木工程，以保证设计与发包安排和施工程序相适应。在施工图阶段，同时提出详细的成本估算。[参见"设计阶段"]

construction document design often "packages" work components, for example, grading and excavation work or foundation and civil engineering work, to assure compatibility of the design with contractual arrangements and the construction process. In this phase, a detailed estimate of costs is also prepared. See also DESIGN PHASE.

Construction documents

Drawings and specifications that define the requirements for a construction project. The construction documents become part of the contract documents. See also CONTRACT DOCUMENTS and WORKING DRAWING.

Construction enterprise

In China, an independent administrative entity, operating with its own budget, that performs construction work. By work type, construction enterprises are classified as general civil engineering companies, companies for machinery and equipment installation, and companies for excavation, foundation construction, or mechanized construction. See also CONSTRUCTION COMPANY.

Construction estimate

In China, calculation of probable construction costs,

施工文件 Shīgōng wénjiàn

规定施工工程要求的图纸和说明书。施工文件是合同文件的组成部分。[参见"合同文件"、"施工图"]

建筑安装企业 Jiànzhù ānzhuāng qǐyè

在中国,指行政上独立、经济上实行独立核算、能担负建筑安装工程施工任务的国营或集体所有制施工单位。按承担工程的性质可分为：从事一般性土木建筑工程的建筑工程公司；从事机器、设备安装的设备安装工程公司；从事专业性土木建筑工程的土石方工程公司、基础工程公司、机械化施工公司等。[参见"建筑公司"]

施工预算 Shīgōng yùsuàn

在中国,指根据施工图和施工定额编制的预算,

prepared from working drawings and construction norms. The construction enterprise uses the construction estimate as the basis for programming a work schedule; for preparing a plan that outlines material and labor requirements; for making work assignments; for implementing brigade accounting and bonus systems; for procuring materials; as well as for implementing management plans and controlling expenditures. See also CONSTRUCTION NORM.

Construction financing

In the United States, procurement of funds to pay for ongoing construction operations. Such funds usually take the form of short-term loans from private-sector lenders that are advanced to the contractor in stages as work progresses and are repaid by the contractor with proceeds from long-term mortgage payments for the completed project. See also DRAWDOWN and GAP FINANCING.

Construction flow process

In China, a construction method involving systematic coordination of all unit processes (structural, electrical, plumbing, and installation work) at various construction levels (interior/exterior, ground/upper floors, floor/ceiling). Coordinating the various processes and closely related simultaneous operations

是施工单位编制施工计划、材料、劳动力计划及对班组下达任务单、实行班组核算、限额领料和推行奖励制度的依据；也是建筑企业对施工加强计划管理、控制成本支出的依据。[参见"施工定额"]

筹措建造资金 Chóucuò jiànzào zījīn

在美国，指为进行中的建筑工程筹措资金。一般是向私营金融界短期贷款，按工程进度分期予付给承包人，承包人用已完工程做抵押。[参见"分期提款"、"资金缺口"]

流水作业法 Liúshuǐ zuòyèfǎ

中国的一种施工组织方法。这种作业法要求在施工过程中，把室内和室外、底层和楼层、地面和顶棚等部分的土建、水电和设备安装等各项工程结合起来，实行多工种、多工序互相穿插，紧密衔接，同时进行，以充分利用空间和争取时间，尽量减少施工中的停歇，加快施工进度。

assures optimal utilization of time and space. The ultimate objective is to minimize work stoppages or delays and to speed up the whole construction process.

Construction for nonproduction purposes

In China, capital construction for directly fulfilling people's living, welfare, and cultural requirements, such as the construction of housing, schools, and hospitals.

Construction for production purposes

In China, capital construction used directly for material production or to meet the requirements of material production, such as the construction of industrial plants, railways, and mines.

Construction headquarters

In China, the coordinated management center operated by the project owner and the design and construction organizations to ensure unified onsite control over execution of the construction project. Often referred to as headquarters.

Construction home base

In China, a semipermanent base established to house family members of mobile construction teams that are moved from one capital construction project

非生产性建设 Fēishēngchǎnxìng jiànshè

在中国,指直接用于人民文化和生活福利需要的基本建设,包括住宅、学校、医院等的建设。

生产性建设 Shēngchǎnxìng jiànshè

在中国,指直接用于物质生产或满足物质生产需要的基本建设,包括工业、运输、地质资源勘探等的建设。

工程指挥部 Gōngchéng zhǐhuībù

由建设单位、施工单位和设计单位共同组成的在施工现场对工程进行统一指挥的联合机构,通常称为指挥部。

生活基地 Shēnghuó jīdì

在中国的基本建设中,一种为流动性施工队伍安置家属的后方半永久性基地,包括住宅及公共福利文化设施。

to another as their services are needed. The base includes residential quarters, as well as health and cultural facilities.

Construction inspector

In the United States, a competent individual appointed by the owner or his representative to check that the work on a construction project conforms to the contractually defined specifications. See also SITE INSPECTION.

Construction management (CM)

Formal advice or coordination services provided on ways to control costs and to speed construction of a building. Construction management emphasizes the time and cost consequences of design construction decisions from bid packaging through project completion. Advanced CM techniques link systems engineering to cost accounting to maximize the cost-effective use of time, labor, and materials, relying on computer-generated schedules to organize on-site construction in great detail. See also COST ACCOUNTING.

Construction management plan

In China, a constituent part of the construction documents that defines the time period, work sequence, and method of construction; the quantity

施工检查员 Shīgōng jiǎncháyuán

在美国，指由业主或其代理人所委派的具有业务能力的人员，负责检查施工项目是否符合合同说明书中所规定的要求。[参见"现场监督"]

施工管理 Shīgōng guǎnlǐ

为控制成本、加速施工进度所提供的正式咨询或协调性任务。施工管理强调在从投标到工程竣工的全过程中，每项设计和施工决策对时间和成本所产生的影响。先进的施工管理技术把系统工程和成本核算结合起来，用计算机编制的进度计划组织现场施工，以便使时间、劳力和材料的利用取得最大经济效益。[参见"成本核算"]

施工组织设计 Shīgōngzǔzhī shèjì

中国施工文件的组成部分。主要内容包括：确定各种工程的施工期限、施工顺序和施工方法，安排各个时期内所需的建筑材料、施工机械和劳动

of building materials, construction equipment, and labor needed at different stages of construction; and an overall plan for the construction site. See also DETAILED CONSTRUCTION SCHEDULE and OVERALL CONSTRUCTION SITE PLAN.

Construction manager

In the United States, an individual who supervises construction of a building, recording materials used and manhours expended. The construction manager is often involved in coordinating and scheduling workers, materials, and subcontractors. See also PROJECT MANAGER.

Construction norm

In China, the usual amount of labor, material, or construction equipment (in machine-shifts, i.e., machine work measured in eight-hour shifts) required by a construction enterprise to accomplish a unit amount of a certain type of construction work, such as excavation, bricklaying, or concrete pouring. The norm is the basis for preparing work element construction programs, for conducting analysis of labor and material requirement, for issuing work assignments, and for evaluating work completed. See also CONSTRUCTION ESTIMATE and NORM.

力数量，拟定和绘制施工总平面图。[参见"施工进度计划"、"施工总平面图"]

施工经理 Shīgōng jīnglǐ

在美国，指监督建筑施工的人员，负责记录人工及材料的消耗，并参与对人工、材料和分包商的协调、联系和进度安排。[参见"工程主任"]

施工定额 Shīgōng dìng'é

中国建筑安装企业为完成某一工种（例如挖土、砌砖、浇灌混凝土等）的单位工程所需要的人工、材料和施工机械台班等数量的标准；是编制施工作业计划、进行工料分析、签发工程任务单和考核生产完成情况的依据。[参见"施工预算"、"定额"]

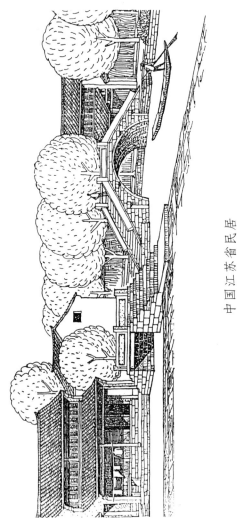

中国江苏省民居

Vernacular dwelling in Jiangsu Province

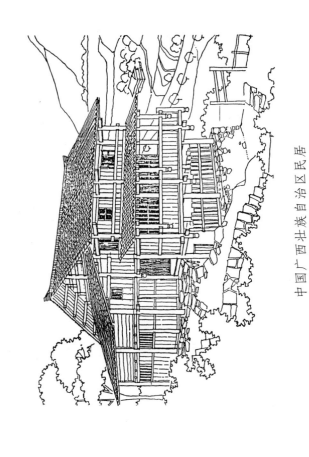

中国广西壮族自治区民居

Vernacular dwelling in Guangxi Autonomous Region

Construction organization

In China, the organization which performs the actual building work for projects. Often referred to in contracts as the second party. See also DESIGN ORGANIZATION and DEVELOPMENT ORGANIZATION.

Construction payment

In China, the total amount to be expended for a construction project, as determined by a detailed estimate. In addition to the construction cost, i.e., direct, indirect, and extra expenses, payment for the completed project includes statutory profit to the construction organization. See also CONSTRUCTION COST and STATUTORY PROFIT.

Construction period

In China, the time expended on the construction of a project or an item of a project, from the start to the completion and acceptance of the project or item.

Construction phase

In the United States, the period of a project devoted to actual construction work.

Construction project

1. In China, a general term for capital construction

施工单位 Shīgōng dānwèi

在中国，指实际承担工程施工的单位。在工程合同中通称为乙方。[参见"设计单位"、"建设单位"]

建筑造价 Jiànzhù zàojià

在中国，指根据设计预算确定的建筑工程的造价，其中包括建筑成本及支付给施工单位的法定利润。建筑成本由直接费、间接费和独立费三部分组成。[参见"建筑成本"、"法定利润"]

施工工期 Shīgōng gōngqī

在中国，指建设项目或单项工程自开始施工到全部建成验收所经历的时间。

施工阶段 Shīgōng jiēduàn

在美国，指工程项目实际施工的那段时间。

建设项目 Jiànshè xiàngmù

1. 在中国，指基本建设项目和更新改造措施的

and renovation projects. The construction project consist of the component items defined in one design program, usually all the construction items of an enterprise or institution, but possibly just a single item. A given project may involve only one or several different sites. See also CAPITAL CONSTRUCTION and CAPITAL CONSTRUCTION PROJECT.

2. In the United States, the term refers to a single structure or a closely related group of structures under construction on one site.

Construction put-in-place

A statistical measure, expressed as a dollar value, representing the total amount of construction activity in the United States (i.e., construction starts, construction in progress, and construction completions) on a monthly and annual basis. Comparison of the figures over a period of years facilitates determination of construction trends. See also HOUSING PRODUCTION.

Construction site facilities

In China, on-site temporary structures and equipment required by a construction enterprise to perform construction work at a new site. Large-scale facilities may include equipment for prefabricating reinforced concrete components

总称。一个建设项目是在一个设计任务书范围内（通常包括一个企业、一个事业单位的全部工程项目，或一个独立工程），在一个或几个场地上进行施工的各个工程项目的总体。[参见"基本建设"、"基本建设项目"]

2. 在美国，指在一个工地上施工的单体建筑物或相互有联系的一组建筑物。

建造量 Jiànzàoliàng

一种以货币价值反映美国施工活动总量的统计指标，包括开工、在施和竣工工程的总量，通常以月度、年度为单位进行统计。将若干年内的该指标加以比较，可以看出建筑业的发展趋势。[参见"住房建造量"]

临时设施 Línshí shèshī

在中国，指建筑安装企业在施工现场临时搭建的必要设施。分为大型临时设施（如混凝土构件预制厂，钢筋、木材加工厂，各种铁路专用线、管道、仓库、办公室，职工生活福利设施）和小型临时设施（如机械、工具库、车棚等）。[参见

or processing timber and reinforcement steel, railroad spurs, pipelines, and warehouses, as well as offices and living facilities for staff and workers. Small-scale facilities encompass such items as machine and tool sheds, and bicycle racks. See also CONSTRUCTION SUPPORT BASE and OVERALL CONSTRUCTION SITE PLAN.

Construction Specifications Institute (CSI)

A private technical society in Alexandria, Virginia, dedicated to the improvement of construction communications, documentation, and specifications through service, education, and research. CSI provides a construction format called MASTERFORMAT which organizes all construction specifications into 16 subject divisions with further subdivisions. The system is commonly used in the United States. See also BUILDING CODES AND STANDARDS.

Construction support base

In China, a semipermanent off-site facility that produces the materials required for a capital construction project. The construction support base usually consists of buildings and equipment for fabricating precast concrete and steel elements, processing lumber, and repairing machines, as well as a sand and stone quarry. See also CONSTRUCTION

"生产基地"、"施工总平面图"]

施工规范研究院 Shīgōng guīfàn yánjiūyuàn

在美国弗吉尼亚州亚历山大的一个私营技术研究机构,专门通过服务、教育和研究致力于改进施工的信息、资料和规范等工作。该机构收集各种施工规范,编成施工专用模式,其中包括16大类,每一类又分成若干章节。这套施工用模式体系在美国被广泛采用。[参见"建筑法规与标准"]

生产基地 Shēngchǎn jīdì

在中国的基本建设中,一种为满足施工需要而建立的各种辅助性的后方半永久性的生产基地。通常由混凝土预制构件厂、木材加工厂、金属构件厂、机械修配厂和沙石采掘场等组成。[参见"临时设施"]

SITE FACILITIES.

Construction team

In China, the basic labor unit in a construction enterprise, with direct responsibility for executing the work processes involved in construction. The construction team is a work organization but not an administrative entity. The unit may be a specialized team, organized to perform a specific function and consisting of workers who practice the same trade, with a certain proportion of ordinary laborers to assist. In contrast, the mixed team is composed of a variety of workers possessing the different skills required to complete the construction work. For the U.S. equivalent, see also CONSTRUCTION CREW.

Construction work element

In China, a component of construction work classified for purposes of cost estimation and construction management according to the main division of a building and the type of installation, e.g., foundations, walls, floors, and pipes. Work may be further subdivided into subelements according to the major trades or purposes, e.g., excavation, concrete work, plastering work, etc. See also ACCEPTANCE OF WORK SUBELEMENTS.

施工队　Shīgōngduì

中国建筑企业内直接组织施工的基层劳动组织。有两种组织形式，即由同一工种的工人配以一定数量的普通工组成的专业施工队；由完成建筑工程所需要的各工种的工人混合组成的混合施工队。[美国的相应词汇参见"施工队伍"]

分部工程　Fēnbù gōngchéng

在中国，指由于预算和施工管理的需要按建筑物的主要部位及安装工程的种类划分的工程，如地基基础工程、墙体工程、地面工程、管道工程等。分部工程又可按主要工作或用途划分为分项工程，如土方工程、混凝土工程、抹灰工程等。[参见"分项工程验收"]

Construction work quantity

In China, a measure of the size of a construction project or work element, expressed in terms of physical units, e.g., building area in square meters, excavation in cubic meters, and pipe installation in linear meters. The figures can be used for cost estimation. See also CONSTRUCTION WORK VALUE.

Construction work subelement

See CONSTRUCTION WORK ELEMENT.

Construction work value

In China, the quantity of construction work expressed in terms of a monetary unit, simply called work value. See also CONSTRUCTION WORK QUANTITY.

Consultant

An outside expert engaged on a temporary basis by a government or private organization to provide specialized technical advice or services. For a construction project, professional consulting services complement or supplement the architect's or engineer's services.

Consulting engineer

See CONSULTANT.

建筑安装工程量 Jiànzhù ānzhuāng gōngchéngliàng

在中国，指用实物单位（如房屋用平方米，土方工程用立方米，管道安装工程用米）来表示的建筑安装工程的量，可用它来估算费用。[参见"建筑安装工作量"]

分项工程 Fēnxiàng gōngchéng

[见"分部工程"]

建筑安装工作量 Jiànzhù ānzhuāng gōngzuòliàng

在中国，指用货币单位来表示的建筑安装工程的量，简称"工作量"。[参见"建筑安装工程量"]

顾问 Gùwèn

政府或私营机构从外单位临时聘请来的专家，为其提供专业技术咨询服务。在建筑工程中，专业咨询工作是对建筑师或工程师工作的一种补充。

顾问工程师 Gùwèn gōngchéngshī

[见"顾问"]

Contingency allowance

In the United States, an amount included in the budget of a project to cover unforeseen costs of labor or materials, or changes in work plan. See also PROJECT BUDGET.

Contract

A written agreement between an owner or governmental department or agency, and the successful bidder or contracting organization covering the performance of work and provision of labor and materials. Under the terms of the legal document, the contractor is bound to perform the work and furnish the labor and materials, and the owner, department, or agency is obligated to compensate him/her at the mutually established and accepted rate in the manner prescribed. See also CHANGE ORDER, CONTRACTOR, GENERAL CONTRACT, OWNER, PRIME CONTRACT, and SUBCONTRACT.

Contract administration

In the United States, supervision of contractually agreed upon activities to ensure that terms of the contract are met. Contract administration for a construction contract may be performed by the architect and engineer or by a special contract officer.

不可预见费 Bùkěyùjiàn fèi

在美国,指工程预算中所包括的一笔款项,用来支付无法预料的人工或材料的费用;也包括因工程计划变更而需要支付的费用。[参见"建设项目投资预算"]

合同 Hétóng

由业主、政府部门或其他代理人与中选投标人或承包单位,为完成工程施工及提供人力与材料等而共同订立的书面协定。这种法律性文件的条款中规定,承包人必须履行条款所规定的施工任务,并提供人工或材料;而业主、政府部门或其他代理人则有责任按规定的方式将议定的价款支付给承包人。[参见"工程变更通知单"、"承包人"、"总承包合同"、"业主"、"直接承包合同"、"分包合同"]

合同管理 Hétóng guǎnlǐ

在美国,指对合同中商定的活动进行监督,以保证合同条款的执行。施工合同的管理,可由建筑师和工程师承担,亦可由专门的合同管理官员负责。

Contract award

Under the U.S. bidding system, notification to a bidder that the submitted bid has been accepted. See also ACCEPTED BID, BID, and BID ABSTRACT.

Contract bond

In the United States, the approved form of security, executed by a subcontractor and his surety or sureties, guaranteeing complete execution of the contract and all supplemental agreements as well as payment of all legal debts pertaining to the construction of the project.

Contract documents

In the United States, any legally binding written materials pertaining to a specific contract. These include the owner-contractor agreement, general and supplementary conditions of the contract, drawings, specifications, all addenda issued prior to execution of the contract, all modifications thereto, and any other items specifically stipulated as being contract documents. See also ADDENDUM, CONSTRUCTION DOCUMENTS, CONTRACTOR'S OPTION, and SPECIFICATIONS.

Contract item

In the United States, an item of work specifically described in a contract, for which a price, either

发包 Fābāo

在美国，指通知投标人其所提出的工程投标已被接纳。[参见"中选标"、"投标"、"标价总表"]

合同担保书 Hétóng dānbǎoshū

在美国，指由承包人或分包人及其保证人共同签具并经过业主同意的保证书，保证完全履行承包合同及其补充协议书，并支付与工程施工有关的一切法定债务。

合同文件 Hétóng wénjiàn

在美国，指业主与承包人所订立的具有法律性的合同，包括合同条款（一般条款、补充条款及其他有关条件）、图纸、说明书以及订立合同前发出的补充或修改文件及其他规定事项。[参见"补充文件"、"施工文件"、"承包人的选择权"、"说明书"]

合同项目 Hétóng xiàngmù

合同中所列出的并附有单价或总价的工作项目。其内容包括全部完成合同中所附说明书或补充说

unit or lump sum, is provided. Work items cover performance of all work and provision of all labor, equipment, and materials stipulated in the text of a specification item of the contract, any subdivision of the text of the supplemental specifications, or special provisions of the contract.

Contractor

Party signing the contract with the owner, organization, or agency to perform the work called for in specifications and/or drawings. In the United States, the contractor may be referred to as a general or prime contractor, depending on the nature of the contractual arrangement. A subcontractor enters an agreement with a contractor to carry out portions of the work. See also CONTRACT, GENERAL CONTRACT, OWNER, and PRIME CONTRACT.

Contractor's option

In the United States, a specific provision of the contract documents which allows the contractor to choose among specified materials, methods, or systems without change in the contract sum. See also CONTRACT DOCUMENTS.

Conurbation

An extensive aggregation or continuous network of cities and towns that has developed spontaneously in

明书中的项目或合同中特殊规定的项目,并提供全部劳动力、设备及材料。

承包人 Chéngbāorén

与业主签订合同按图纸及说明书承建工程的一方。在美国,根据合同性质,有总承包人或直接承包人。与承包人签订合同承建部分工程者为分包人。[参见"合同"、"总承包合同"、"业主"、"直接承包合同"]

承包人的选择权 Chéngbāorénde xuǎnzéquán

在美国,合同文件中规定承包人可自行选择几种指定的材料、方法或体系,而不改变合同总金额。[参见"合同文件"]

集合城市 Jíhé chéngshì

在一定的地理范围内由多个自发建设的城市构成的聚集体。

a given geographical sector.

Convenience outlet

See CONVENIENCE STORE.

Convenience store

1. In the United States, a readily accessible, usually small retail business that sells frequently purchased articles for immediate use, for example, newspapers and cigarettes.
2. In the United States, a large retail business that sells a variety of high-demand consumer goods at reasonable prices. Merchandise is arranged on shelves along aisles to permit self-service and ease of shopping. Also called a convenience outlet.

Conventional construction

Traditional designs, methods, and materials for building in a given area. In colloquial American usage, conventional construction designates wood-frame, single-family houses.

Core housing

Housing units furnished with only the basic structure complete to reduce construction costs. The remainder of the work can be finished by the new owners on their own time and at their own expense.

便利商店 Biànlì shāngdiàn

[见"杂货店"、"自选商店"]

1. 杂货店 Záhuòdiàn

在美国,地点靠近的、出售日常用品(如报纸、香烟等)的零售小店。

2. 自选商店 Zìxuǎn shāngdiàn

在美国,出售多种销售量大而价格便宜的消费品的大型零售店。货物陈列在货架上,由顾客自取选购。又称作"便利商店"。

传统建筑 Chuántǒng jiànzhù

指一定地区内按传统的设计方法和材料进行建造的建筑。在美国俗语中,一般指木构架的独户住宅。

核心住房 Héxīn zhùfáng

为节约造价,提供仅有基本结构的房屋,其余部分可由购房人自费随时添建。此种住房流行于发展中国家。[参见"自建住房"、"提供场地与基础设施"]

Housing of this type is popular in developing countries. See also SELF-HELP HOUSING and SITE AND SERVICES.

Corporation

In the United States, legal identity consisting of employers and employees organized privately to pursue business activities of a particular type or types.

Cost accounting

The branch of accounting concerned with analyzing, classifying, and recording the cost elements of material, labor, and other expenses involved in producing a good or rendering a service. See also CONSTRUCTION MANAGEMENT.

Cost-benefit analysis

Quantitative examination of the potential benefits and costs from various alternative prospective systems to identify the preferred systems, equipment, and products.

Cost breakdown

A detailed enumeration of the costs of the separate items constituting a total budget. The objective is to identify which costs might be eliminated or reduced.

公司 Gōngsī

在美国，指由雇主及雇员组成的一种私营的、具有法人资格的实体，从事一种或几种商业活动。

成本核算 Chéngběn hésuàn

会计工作的一部分。其内容是：对生产一种产品或提供一种服务所需的材料、人工及其他费用等各项成本进行分析、分类、记录。[参见"施工管理"]

成本效益分析 Chéngběn xiàoyì fēnxī

对可供选用的不同体系的潜在效益和成本所作的定量分析。其目的是判明较优的体系、设备或产品。

成本分类 Chéngběn fēnlèi

详细列出总预算中各个项目的成本，以确定哪些成本可以取消或缩减。

Cost controls

In the United States, measures taken by management after cost analysis has been conducted to limit expenditures for a project or organization and to use funds efficiently.

Cost-effective

Producing the desired results at the least cost.

Cost estimate

An approximation of expenditures required to carry out a particular project, produce a good, or render a service.

Cost overrun

In the United States, the amount of money expended over and above budgeted costs for any project, including such items as labor, interest, materials, and land.

Cost per square foot/meter

See UNIT COST.

Cost-plus-fee contract

In the United States, an agreement in which the contracting party agrees to reimburse the contractor (or architect-engineer) for direct and indirect costs, plus an additional fixed fee for services, usually a

成本控制 Chéngběn kòngzhì

在美国,指管理部门在进行成本分析后所采取的措施,以便限制某一计划或组织的开支,并有效地使用资金。

良好经济效益 Liánghǎo jīngjì xiàoyì

以最低的成本取得理想的效果。

成本估计 Chéngběn gūjì

为完成一项工程、生产一种产品或提供一项服务所需费用的概略估计。

成本超支 Chéngběn chāozhī

在美国,指超过预算成本的支出金额,包括用于人工、利息、材料和土地的款项。

单方成本 Dānfāng chéngběn

[见"单位成本"]

成本加费用合同 Chéngběn jiā fèiyòng hétóng

在美国,指业主与工程承包人签订的一种协议。业主偿付给承包人(或建筑师—工程师)的工程价款为承包人所花费的直接费和间接费,外加规定数额或按上述费用的一定百分比计算的服务费。

stipulated sum or a percentage of the costs.

Cost-price index

A statistical measure of the relationship between the cost of producing a product and the offering price of that product. See also UNIT COST and UNIT PRICE.

Cottage industry

A type of business in which goods are produced or processed by families in their homes and marketed either by the families themselves or by a merchant employer. See also INDIVIDUAL BUSINESS, NEIGHBORHOOD INDUSTRY, and SIDELINE PRODUCTION.

Council of Governments (COG)

In the United States, a voluntary coordinating body composed of elected officials from different local governments in a metropolitan area or rural region. These officials meet regularly to coordinate regional planning and to organize other types of cooperative activities.

County

1. In China, the local administrative subdivision of a province, an autonomous region or prefecture, or of a centrally or provincially administered municipality.

成本价格指数 Chéngběn jiàgé zhǐshù

一种统计指数,用以表示生产某一产品的成本与其售价之间的关系。[参见"单位成本"、"单价"]

家庭工业 Jiātíng gōngyè

在家庭中制造或加工产品的工业。产品由家庭自销或由商人代销。[参见"个体户"、"街道工厂"、"副业生产"]

规划协调委员会 Guīhuà xiétiáo wěiyuánhuì

在美国的大都市地区或农村地区,由各地方政府选举的官员组成的一种机构,定期开会并根据自愿原则协调区域规划及其他合作事项。

县 Xiàn

1. 在中国,省、自治区、直辖市和自治州或省辖市以下的行政区域。城市远郊地区由市管辖的县,称市辖县,设县政府。行政上由省或自治区管辖

The county government may be administered by the municipality to which it is adjacent, or may be under the jurisdiction of a province or autonomous region, with only an economic affiliation to the nearby municipality. See also SUBURB.

2. In the United States, the primary administrative subdivision of the state. Counties are called "parishes" in Louisiana.

County manager

See CITY MANAGER.

Courtyard house

A dwelling type generally adopted in North China in the Ming (1368-1644 A.D.) and Qing(1644-1911 A.D.)Dynasties, most commonly found in Beijing even to the present day. The house is laid out according to the concept of the patriarchal clan system and etiquette in the feudal period of China. The rooms and courtyard are symmetrically arranged along the south-north axis. The main rooms facing south are for senior members of the family, whereas the rooms in the wings facing east and west are for younger generations. The rooms are connected with a winding corridor or a veranda. Side-rooms with small yards, usually attached to both sides of the main rooms, contain the kitchen,

而经济上由市管理的非市辖县,称市管县。[参见"郊区"]

2. 在美国,指州以下的最高行政区域,但在路易斯安那州则称"教区"。

县行政官 Xiàn xíngzhèngguān

[见"市行政官"]

四合院 Sìhéyuàn

中国明(1368—1644)、清(1644—1911)两代北方一般城镇住宅的一种典型形式,以北京最为普遍,至今尚存在相当数量。这种住宅的布局,以封建宗法礼教为指导思想,按南北纵轴线对称地布置房屋和院落。朝南正房供长辈居住,东西厢房是晚辈的住处,周围用走廊联系。正房的左右,常附以耳房与小跨院,置厨房、杂屋和厕所。住宅四周由各座房屋的后墙及围墙所封闭,一般不对外开窗。窗、门均朝中间的院子开设。院内栽植花木,亦有设荷花池、陈设盆景等。[参见"民居"]

toilet, and storerooms. The courtyard is usually enclosed by rooms on all four sides. There are no exterior windows in the outer walls; instead, all doors and windows face the courtyard, which is planted with flowers and trees, and sometimes decorated with miniature trees and lotus ponds. See also "VERNACULAR HOUSING."

CPM

See CRITICAL PATH METHOD.

Critical Path Method (CPM)

A technique that serves as a management tool for planning and scheduling projects. An arrow diagram is drawn to show the interconnected individual tasks required to complete the project. The diagram permits determination of the relative significance of each event and establishment of the critical path, i.e., the optimal sequence and duration of operations. See also PROGRAM EVALUATION AND REVIEW TECHNIQUE.

Crop loss compensation

In China, remuneration paid by a development organization to commune members or a production brigade for the loss of the productive crop capacity of farmland that has been requisitioned with the

关键线路法(缩写) Guānjiàn xiànlùfǎ (suōxiě)

[见"关键线路法"]

关键线路法 Guānjiàn xiànlùfǎ

用于工程计划和进度安排的一种管理手段。用箭头图表示完成工程所需的有内部联系的各工种项目。从图表中可以确定每一事项的相对重要性,从而得出关键线路,即最优顺序及操作时间。[参见"计划评审法"]

农田青苗补偿 Nóngtián qīngmiáo bǔcháng

中国建设单位在城市规划部门批准征用农田后,付给在该农田上生产的社员或生产队青苗损失的赔偿费。[参见"征地"]

approval of the urban planning administration. See also LAND ACQUISITION.

CSI

See CONSTRUCTION SPECIFICATIONS INSTITUTE.

Cul-de-sac

In the United States, a short dead-end street with no through-traffic and limited on-street parking. In planned residential areas, the cul-de-sac usually ends in a traffic circle. See also DEAD END.

Cultural district

See DISTRICT.

Culturally sensitive design

In the United States, planning of architectural or site characteristics, as well as space and land use to accommodate the cultural values and preferences of a specific local ethnic or social group, or time period.

施工规范研究院（缩写） Shīgōng guīfàn yánjiūyuàn (suōxiě)

[见"施工规范研究院"]

尽端路 Jìnduānlù

在美国，一端不通、不能穿行且限制停车的短路。在按规划建成的居住区内，路的尽端常呈环状。[参见"死胡同"]

文教区 Wénjiàoqū

[见"区"]

文化特色区设计 Wénhuà tèsèqū shèjì

在美国，为考虑保存具有文化价值和某一地区种族、社会集团或历史时期的特殊需要，在规划中对建筑或地段特点以及空间和土地的利用所作的特殊设计。

D

Day-care center

A special facility that provides care and supervision for infants and preschool children during the day while their parents are at work. In the United States, day-care centers also exist for elderly and handicapped persons unable to care for themselves.

Daytime population

The people present in a given area during the day, usually to work. A central business district may have a large daytime population but be almost deserted at night.

Dead end

A small street, often in a residential area, that ends abruptly without an outlet. See also CUL-DE-SAC.

Debenture

In the United States, a bond issued by a government or corporation without security, or an obligation not secured by a specific lien on property. This form of long-term debt is backed only by the assets and credit of the issuing organization. See also BOND.

Debt service

The regular payment and collection of principle and interest due on a loan or mortgage.

日托中心 Rìtuō zhōngxīn

一种在日间看管双职工的幼儿以及学前儿童的设施。在美国,也有照顾不能自理的老年人及残疾人的日托中心。

昼间人口 Zhòujiān rénkǒu

一定区域内昼间的人口数,通常在该区域工作的人,有些商业中心区的昼间人口可能很大,但夜间几乎空无一人。

死胡同 Sǐhútòng

居住区内无出口的小巷。[参见"尽端路"]

信用债券 Xìnyòng zhàiquàn

在美国,指由政府或公司发行的一种债券,这种债券无须担保品,没有应尽义务,也没有以指定的财产留置权作为保证。此种长期债券仅以发行机构的财产和信用作为依靠。[参见"债券"]

债务清偿服务 Zhàiwù qīngcháng fúwù

定期收复贷款或抵押的本、息。

Decentralization

1. Delegation of governmental functions and policy-making powers from a central authority to dispersed units, e.g., national to regional or state to local. The opposite concept, centralization, entails concentration of power in a central authority.

2. Spontaneous or planned movement of people or industries from the center of an urban area to the fringes or to separate towns or satellite cities outside the city proper. Decentralization may also apply to the dispersion of an industry from one region to a number of regions.

Decision making process

The logical procedures for evaluating options and choosing among them, especially regarding policies in government or business.

Declining neighborhood

In the United States, a neighborhood, frequently adjacent to the city center, characterized by poorly maintained housing and a run-down appearance. Deterioration is generally associated with a decline in property values, a simultaneous influx of low-income disadvantaged residents, and an increased incidence of crime. See also TRANSITIONAL NEIGHBORHOOD.

权力分散　Quánlì fēnsàn

1. 将政府的职能及决策权自中央权力机构分散到下属单位。例如从国家分散到地区，从州分散到地方。相反的概念是集中化，即将权力集中在中央机构。

2. 人口或工业自发地或有计划地由市区中心向边缘的城镇或卫星城市转移。亦可指一种工业从一个区域分散到几个区域。

决策过程　Juécè guòchéng

对几种可能性的评价并加以选择的逻辑过程，尤指有关政府或企业界的决策。

衰退邻里　Shuāituì línlǐ

在美国，指住房陈旧失修、面貌颓败的邻里，常位于市中心附近。住房的陈旧导致房产贬值、生活条件差的低收入住户随之迁入以及犯罪率上升。[参见"演变中的邻里"]

Default

Failure of a party to an agreement to meet his/her contractual obligations under that agreement, which may be a contract, a lease, or a mortgage. See also DELINQUENT MORTAGAGE and FORECLOSURE.

Defensible space

A theory popularized by the architect Oscar Newman (*Defensible Space: People and Design in the Violent City,* 1972) which holds that such factors as the layout of neighborhoods, the form of streets, the design of housing and the planning of schools in a city can contribute to the occurrence of crime. According to this view, crime can be reduced by the modification of the features of the environment that are conducive to crime.

Deferred project

In China, one of the main statistical indices of capital construction. The term refers to construction projects halted or delayed by the responsible authorities in accordance with the national economic readjustment policy or cutbacks in capital construction.

Delinquent mortgage

In the United States, a mortgage loan for which the

违约 Wéiyuē

订立契约（合同、租约或抵押贷款）的一方未按契约规定履行其所承担的义务。[参见"拖欠抵押贷款"、"取消赎回权"]

防御性空间 Fángyùxìng kōngjiān

由建筑师奥斯卡·纽曼提出的一种理论（见《防御性空间：暴力城市中的人民与设计》，1972年），认为城市中邻里的布置、街道的形式、住宅的设计、学校的规划等因素，如不适当，将促使犯罪行为的发生。根据这一理论，改进环境中足以引起犯罪事件的特点可以降低犯罪率。

停建缓建项目 Tíngjiàn huǎnjiàn xiàngmù

在中国，基本建设主要设计指标之一。根据国民经济调整方针或压缩基本建设规模的要求，经有关部门批准，停止施工与暂缓建设的项目。

拖欠抵押贷款 Tuōqiàn dǐyā dàikuǎn

在美国，指拖延偿付抵押贷款。如继续拖欠则将

payments are overdue. Further delay in payment may result in legal action against the borrower. See also DEFAULT, FORECLOSURE, and MORTGAGE.

Demography

The study of human population statistics and characteristics, including factors such as population distribution and size, growth trends, and social data (e.g., age, race, ethnic background, etc.) in a given country, region, or city. See also AGE DISTRIBUTION.

Demonstration project

See PILOT PROJECT.

Density of living floor area

In China, the percentage of total floor area of living rooms and bedrooms in dwelling units as compared to the total site area in a residential district, subdistrict, or housing group. See also LIVING FLOOR AREA and RESIDENTIAL DENSITY.

Density of registered inhabitants

In China, the number of people with resident registration per unit area in a residential district, subdistrict, or housing group. See also RESIDENTIAL DENSITY.

导致对借款人采取法律行动。[参见"违约"、"取消赎回权"、"抵押贷款"]

人口统计学 Rénkǒu tǒngjìxué

关于人口统计及特征的研究,包括在一特定国家、地区或城市内人口分布、数量、增长趋势及各种社会调查数据(如年龄、民族、种族背景等)。[参见"年龄分布"]

示范工程 Shìfàn gōngchéng

[见"试点工程"]

居住面积密度 Jūzhù miànjī mìdù

在中国,指居住单元内居室面积总和与居住区、居住小区或住宅组团内土地总面积之比。[参见"居住面积"、"居住密度"]

居住人口密度 Jūzhù rénkǒu mìdù

在中国,指居住区、居住小区或住宅组团中,单位面积内持有正式户口的居住人口数。[参见"居住密度"]

Density of residential floor area

In China, the percentage of total floor area of residential buildings as compared to the total site area in a residential district, subdistrict, or housing group. See also RESIDENTIAL DENSITY and RESIDENTIAL FLOOR AREA.

Density transfer

In the United States, a technique for residential zoning in which an average density is established for the whole community and high-density development areas are offset by large open spaces. See also CLUSTER ZONING and TRANSFERABLE DEVELOPMENT RIGHTS.

Dependent population

The portion of the total population of an area that does not belong to the working population, including minors, the disabled, and individuals not employed. In China, also called the nonworking population.

Depopulation

Reduction of the number of inhabitants of an area, usually as the result of one or more economic, social, environmental, or government policy measures.

Deposit for bidding documents

In the United States, monetary deposit required

居住建筑面积密度 Jūzhù jiànzhù miànjī mìdù

在中国,指居住区、居住小区或住宅组团内,居住建筑的总面积与土地总面积之比。[参见"居住密度"、"居住建筑面积"]

密度调剂 Mìdù tiáojì

在美国,对居住区进行区划的一种方法,即定出整个区域的平均建筑密度,对高密度的建设区用大面积的室外空间加以调剂。[参见"组团式区划"、"可转让开发权"]

被抚养人口 Bèifǔyǎng rénkǒu

城市总人口中除去劳动人口外的人口,指未成年的、没有劳动能力的以及未参加劳动的人口。在中国,亦称非劳动人口。

人口减少 Rénkǒu jiǎnshǎo

指某一地区人口减少的现象,常由一种或几种经济、社会、环境或政策等因素所引起。

投标文件押金 Tóubiāo wénjiàn yājīn

在美国,投标人领取施工文件和投标须知时所付

to obtain a set of construction documents and bidding requirements, customarily refunded to bona fide bidders on return of the documents in good condition within a specified time. See also BIDDING DOCUMENTS.

Depreciation

A loss in value of real property resulting from age, physical deterioration, obsolescence, or some other cause. In the United States, a depreciation amount may be deducted from the taxable income of the owner of investment property. See also ACCELERATED DEPRECIATION and APPRECIATION.

Depressed area

A geographic region in which unemployment is high and economic conditions are stagnant.

Description of construction progress

In the Chinese construction management system, a descriptive evaluation of the current stage and degree of completion of construction work, expressed in terms of the particular portion or percentage of work completed. Required at regular intervals, such descriptions are used in statistical indices and as part of a system for tracking project progress.

的现款押金,通常对守信的投标人在按规定时间内完好无损地送回文件时即予返还。[参见"招标文件"]

折旧、贬值 Zhéjiù、biǎnzhí

房地产因年久、自然损坏、过时或其他原因而降低价值。在美国,房地产的业主可以在其应交所得税的收入中扣除折旧费。[参见"加速折旧"、"增值"]

不景气地区 Bùjǐngqì dìqū

指失业率高、经济停滞的地理区域。

工程形象进度 Gōngchéng xíngxiàng jìndù

中国的施工管理体制中,对建筑安装工程中已完工的具体部位的文字叙述或实物工程量百分比。为定期反映工程进度和计划完成情况的统计方法。

Desegregation

The process of removing legal and social barriers that separate races within a society and its institutions, especially in schools and residential areas.

Design brief

See DESIGN PROGRAM.

Design-build process

In the United states, an approach in which one individual or organization agrees contractually to complete both design and construction of a structure. Design and construction are usually undertaken by two separate organizations in China. See also ARCHITECTURAL ENGINEERING FIRM, PACKAGE DEAL, and TURNKEY PROJECT.

Design development

1. In China, the intermediate phase when design work is conducted in three phases. Design development entails coordination of various design elements and specifications, as well as revision of preliminary estimates. When approved, the modified plans and revised preliminary estimates serve as the basis for preparation of working drawings and for appropriation of capital construction funds.

2.In the United States, the intermediate phase of

消除种族隔离 Xiāochú zhǒngzú gélí

在社会及其团体中,消除法律上和社会上所造成的种族隔离障碍,尤其是在学校和居住区中。

设计概要 Shèjì gàiyào

[见"设计任务书"]

设计兼施工 Shèjì jiān shīgōng

在美国,指由同一机构或个人兼行承包建筑物的设计和施工的做法。在中国,设计和施工常由两个机构分别承担。[参见"建筑师工程师事务所"、"一揽子交易"、"交钥匙工程"]

技术设计 Jìshù shèjì

1. 在中国,指设计工作按三阶段进行时的中间阶段。其基本内容是根据批准的初步设计文件进行编制,协调各有关专业的图纸、说明书,并修正概算。经送审批准的技术设计和修正总概算是建筑工程拨款和编制施工图等设计文件的依据。

2. 在美国,指建筑设计的中间阶段。在此阶段,根

architectural design in which schematic design documents are modified according to the wishes of the client. Design development documents, including double-line drawings with such details as window depth and electrical conduits, define the project's architectural, structural, mechanical, and electrical systems, as well as fairly detailed specifications on materials to be used and long-term procurement items that must be ordered well in advance of construction. Drawings and specifications are sufficiently detailed at this stage that they could be used for actual construction of a turnkey project, and cost estimates are refined to within five to ten percent of the final actual costs. See also DESIGN PHASE.

Design organization

In China, an organization commissioned by a development organization (i.e., the construction project owner) to perform design work in accordance with the project scope and the design program; sometimes referred to contractually as the third party. The organization responsible for architectural design is generally a design institute. See also CONSTRUCTION ORGANIZATION and DEVELOPMENT ORGANIZATION.

据委托人的要求修订方案设计文件。技术设计文件包括标明诸如窗户高度和电气导管等细节的双线图，确定工程项目的建筑、结构、机械、电气系统以及所使用材料的详细规格，订货期较长的项目必须在施工前及早订货。技术设计阶段提出的图纸和确定的规格须详细到足以满足交钥匙工程实际施工的需要，所估算的成本需精确到与最终实际成本仅相差5%—10%。[参见"设计阶段"]

设计单位　Shèjì dānwèi

在中国，指受建设单位委托，按建设任务和设计任务书的要求进行设计的单位。采用承包方式时，亦称"丙方"。在中国，主要的设计工作是由设计院承担的。[参见"施工单位"、"建设单位"]

Design phase

Stages of architectural design work for a construction project. In China, design work is usually carried out in two phases, i.e., preliminary design and construction document design. For large and complicated projects, an intermediate phase, called design development, is added between the two. In the United States, architectural design encompasses three phases: schematic design, design development, and construction document design. See also CONSTRUCTION DOCUMENT DESIGN, DESIGN DEVELOPMENT, PRELIMINARY DESIGN, and SCHEMATIC DESIGN.

Design program

A written description specifying design objectives and constraints of a project, including the size and distribution of particular area types and special system and equipment requirements. In the United States, the program is drafted prior to actual design by the architect working closely with the client. In China, the design organization proceeds with the design phase according to requirements set out in the program prepared by the development organization (the client).The proposed project may encompass one item, e.g., a single building, or a large number of items that will be regarded collectively as one

设计阶段　Shèjì jiēduàn

指建筑工程建设中设计工作的阶段。在中国的一般工程中分为初步设计和施工图设计两个阶段。在大型、复杂的工程中分为初步设计、技术设计和施工图设计三个阶段。美国的建筑设计分为方案设计、技术设计和施工图设计三个阶段。[参见"施工图设计"、"技术设计"、"初步设计"、"方案设计"]

设计任务书　Shèjì rènwùshū

说明一个工程项目的设计要求及条件的书面文件,包括各种不同用途的面积的分布、各种体系及设备的要求等。在美国,设计任务书是在具体设计工作开始前由建筑师根据业主要求制订的。在中国,设计任务书是在具体设计工作开始前由建设单位(即业主)制订的,由设计单位根据其要求进行设计。包括在一个设计任务书内的工程项目称为一个建设项目,可以是一个单项工程,也可以是一组建筑物。在英国,称为"设计概要"。[参见"设计阶段"]

construction project. In Great Britain, called a design brief. See also PROGRAMMING PHASE.

"Design with Nature"

An ecological approach to regional planning developed by Ian McHarg (*Design with Nature*, 1969). The approach focuses on the capacity of an area to absorb development and uses overlay maps to analyze the compatibility of various ecosystems in the region with broad categories of development such as residential, commercial, industrial, and recreational. See also CARRYING CAPACITY, ENVIRONMENTAL DESIGN, LANDSCAPE ARCHITECTURE, and URBAN ECOLOGY.

Designated function of a city

In China, the part a city plays in politics, economy, and culture either in the State or in a region. The function is defined in the comprehensive planning of the city based on the dominant factors in its formation and development.

Detached house

A single-family house that stands apart from others and is surrounded by land on all sides. Also called a single-family detached house. See also ATTACHED HOUSE and HOUSING PREFERENCE.

适应自然的设计 Shìyìngzìránde shèjì

一种区域规划的生态学观点,由麦克哈格在《适应自然的设计》(1969年)一书中所提出,强调一个区域在开发上的吸收能力,并用叠加的地图分析各种生态系统共存的和谐性。区域内的各种开发大致分为居住、商业、工业、娱乐等。[参见"持续能力"、"环境设计"、"风景建筑学"、"城市生态学"]

城市性质 Chéngshì xìngzhì

在中国城市的总体规划中,根据城市的形成与发展的主导因素确定他在国家和地区的政治、经济、文化中的地位和作用。

独立式住宅 Dúlìshì zhùzhái

周围均有空地的独户住宅,一般与其他建筑不相连。[参见"毗连住宅"、"住房选择"]

Detailed construction schedule

In China, a major constituent part of a construction management plan. The main contents are the milestones for the completion in sequencing of construction work elements, for example, the sequence of the erection of form-work, securing of reinforcements, and pouring of concrete, as well as the time needed for each step in the sequence. See also CONSTRUCTION MANAGEMENT PLAN.

Detailed estimate

1. In China, calculation of probable costs, prepared on the basis of working drawings for a particular project, detailed estimate norms, unit costs, and service fee norms. The detailed estimate is used as the basis for determining construction costs, signing contracts and agreements, cost settlement between the project owner and the construction organization, and calculation of work quantity for programs and statistics. Also called a working drawing estimate. See also NORM FOR DETAILED ESTIMATES.

2. In the United States, a forecast of construction costs prepared on the basis of a detailed analysis of materials and labor for all items of work, as contrasted with an estimate based on current area, volume, or similar unit costs. See also

施工进度计划　Shīgōng jìndù jìhuà

中国施工组织设计的重要组成部分,主要内容为确定建筑安装工程的施工期限和施工顺序。例如:支模板、绑钢筋、浇灌混凝土等工序的施工顺序及期限。[参见"施工组织设计"]

设计预算　Shèjì yùsuàn

1. 在中国,指根据某一建筑安装工程的施工图和预算定额、预算单价及取费标准编制出来的该工程的预算。用设计预算确定的预算价格是确定工程造价的依据,是建设单位与施工单位签订合同、协议、结算工程价款的依据,也是计划统计工作中计算建筑安装工作量的依据。又称作"施工图预算"。[参见"预算定额"]

2. 在美国,指按工程项目详细估算工程造价。是根据全部工程项目所用工料的详细分析做出的,不同于根据现有的以面积、体积或类似单价做出的预算。[参见"建筑成本"]

CONSTRUCTION COST.

Detailed planning

In China, planning of an urban area to be developed toward short-term goals with respect to the disposition of buildings and structures, public utilities, and green spaces. Detailed planning defines a course for physical implementation of comprehensive planning, and in turn provides the basis for architectural designs, including the selection of technical-economic indices, definition of space treatment requirements in architecture and determination of coordinates of the construction sites and building locations and levels. The result of detailed planning is a detailed plan. See also COMPREHENSIVE PLANNING and URBAN PLANNING.

Deterioration

Decline in value of a property or neighborhood resulting from wear, weather, abuse, or lack of maintenance.

Developer

In the United States, an individual, company, or corporation engaged in the development and improvement of land for construction purposes, as well as the construction of buildings on that land.

详细规划　Xiángxì guīhuà

在中国，指按城市总体规划的要求，对城市局部地区近期需要建设的房屋建筑、市政工程、园林绿化等做出具体布置的规划，为建筑设计提供依据。内容包括：选定技术经济指标，提出建筑艺术处理要求，确定各项用地的控制性坐标与标高等。[参见"总体规划"、"城市规划"]

破落　Pòluò

建筑物或居住小区因久用、失修或滥用而造成产业的贬值。

开发者，建设者　Kāifāzhě, jiànshèzhě

在美国，指从事建设用地的开发和改进，并在该土地上进行建造活动的个人、公司或企业。

Federal Style, 1780-1820
Georgetown rowhouses
Washington, D.C.
19世纪初期联立式住宅

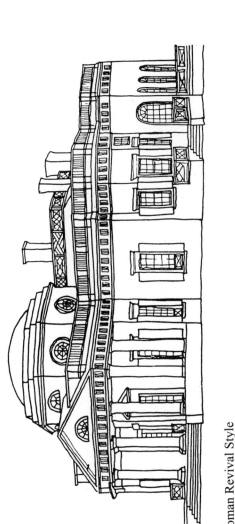

Roman Revival Style
Monticello, 1772-1779
Charlottesville, Virginia
(Architect: Thomas Jefferson)

杰斐逊总统的罗马复兴式住宅

Development

The performance of building, engineering, mining, or other operations in, on, over, or under land, or material change in the use of any buildings to improve the utility of land or property.

Development organization

In China, the organization that wishes to have a project developed. Also called the project owner, or in contracts, the first party. See also CONSTRUCTION ORGANIZATION and DESIGN ORGANIZATION.

Development rights

See TRANSFERABLE DEVELOPMENT RIGHTS.

Dilapidated housing

Housing that does not provide safe and adequate shelter and in its present condition endangers the health, safety, or well-being of the occupants, according to local standards and inspection procedures. See also SUBSTANDARD HOUSING and TENEMENT BUILDING.

Dimensional coordination

Establishment and acceptance of size standards for building elements such as windows and doors so that suitable preconstructed units can be readily

开发,建设 Kāifā, jiànshè

在地上、地下或空中进行的建造、工程建设或其他活动,或对任何房屋所做的物质上的改变,以改进对土地或产业的利用。

建设单位 Jiànshè dānwèi

在中国,指需要建设一项工程的单位,亦称"业主"。采用承包方式时,通称"甲方"。[参见"施工单位"、"设计单位"]

开发权 Kāifāquán

[见"可转让开发权"]

危房 Wēifáng

指根据地方标准及检验程序,被确定为不足以避风雨,其现状将危及住户的卫生、安全或健康的住房。[参见"不合标准住房"、"经济公寓"]

尺寸协调 Chǐcùn xiétiáo

指确定与采用建筑构、配件如门、窗等的标准尺寸,以便于购买到适用的预制部件。[参见"专用建筑体系"、"模数协调"、"通用建筑体系"]

obtained. See also CLOSED SYSTEM, MODULAR COORDINATION, and OPEN SYSTEM.

Direct expenses

Expenses directly incurred and assignable to the account of a particular project, including labor and material costs, expenses for the use of equipment, and miscellaneous direct expenses, e.g., fees for electricity and water use, special measures to permit construction in winter, and repeated transportation of building materials. See also CONSTRUCTION COST, EXTRA EXPENSES, and INDIRECT EXPENSES.

Directly related family

See FAMILY.

Disaster planning

Urban planning that considers precautionary measures to protect human life and property against natural calamities such as earthquakes or windstorms. See also EMERGENCY HOUSING and SEISMIC PLANNING.

Discretionary funding

Grant of funds to an eligible local government for technical assistance and community develop-

直接费 Zhíjiēfèi

指能够并适宜于直接计入工程项目的费用,包括直接用于工程的人工费、材料费、施工机械使用费和其他直接费,诸如工程用水电费、冬雨季施工费、材料二次倒运费等。[参见"建筑成本"、"独立费"、"间接费"]

直系家庭 Zhíxì jiātíng

[见"家庭"]

城市防灾规划 Chéngshì fángzāi guīhuà

城市规划中抵御地震、洪水、风灾等自然灾害,为保护人类生命财产而采取预防措施的规划的通称。[参见"应急住房"、"城市抗震规划"]

选择性拨款 Xuǎnzéxìng bōkuǎn

由美国住房与城市发展部部长决定,拨给符合条件的地方政府,作为技术资助社区建设用的资金。

ment at the discretion of the Secretary of the U.S. Department of Housing and Urban Development. Such funds are available under the Title I – Housing and Community Development Act of 1974. See also COMPETITIVE FUNDING and FORMULA FUNDING.

Displacement of families

In the United States, forced or voluntary relocation of families because of public land acquisition programs, renewal projects, shifts in industrial or commercial activity, or natural disasters. See also JOB DISPLACEMENT and RELOCATION.

Distressed city

In the United States, a deteriorating urban area with a stagnant business climate and joblessness, high crime rates, and inadequate or poor public services, Official classification of a city as "distressed" qualifies it for block grant funds. See also ENTITLEMENT CITY.

District

1.Part of a city having its own specific character, resulting from agglomerated elements which define a certain economic, social, and administrative unity. In China, a city comprises several districts, which

这种资金系根据美国1974年《住房和社区发展法》第一条的规定拨给。[参见"竞争性拨款"、"按公式拨款"]

家庭迁移 Jiātíng qiānyí

在美国,由于公用土地征用计划、更新计划、工业或商业活动的转移、自然灾害等原因引起的自愿的或被迫的家庭搬迁。[参见"就业转移"、"搬迁"]

贫困城市 Pínkùn chéngshì

在美国,指经济停滞、失业严重、犯罪率高、公共服务设施不足或不良的城市地区。经正式宣布为"贫困"的城市可获得政府的一揽子赠款。[参见"受资助城市"]

区 Qū

1. 指城市内具有不同特征的地域范围,在社会、经济和行政上形成一个整体。在中国,城市由若干个市区组成,市区设区政府,由市政府领导。市区由若干街道组成,街道设街道办事处。[参见

in turn encompass a number of subdistricts. The administrative authority of a district is called a district government and is under the jurisdiction of the municipal government. The administrative body of the subdistrict is called the subdistrict office. See also SUBDISTRICT OFFICE.

2. District can also refer to an area in a city where development of a certain type is concentrated, for example, an industrial district, a commercial district, a cultural district, or a residential district. District in this sense is not an administrative entity. In the United States, district types may be differentiated officially by zoning restrictions, and fixed boundaries may be set for certain special-purpose districts, e.g., voting districts. See also CENTRAL BUSINESS DISTRICT, CITY DISTRICT PLANNING, HISTORIC DISTRICT, RESIDENTIAL DISTRICT, SECTORIZATION OF FUNCTIONS, SPECIAL DISTRICT, and ZONING DISTRICT.

District heating

A method used in China for providing space heating to buildings, houses, apartment houses, offices, and industrial premises in an urban area by conducting steam from a central source. It is also used in the United States, but on a very limited basis.

"街道办事处"]

2.指城市内特有的活动区域,或某种机构集中设置的地方,例如工厂集中的地方称工业区,依此类推,有商业区、文教区、居住区等。这种区一般不具有行政职能。在美国,亦可由官方根据区划规定而确定不同性质的区,因某种特殊需要而设置的区还可有固定的界限,如选举区等。[参见"商业中心区"、"城市分区规划"、"历史性市区"、"居住区"、"功能分区"、"特区"、"区划市区"]

区域供热 Qūyù gōngrè

中国城市地区采用的一种由中心热源向住宅、办公楼、工业用房等供热的方法。在美国也有,但仅在有限的范围内采用。

District housing management bureau

See MUNICIPAL HOUSING MANAGEMENT BUREAU.

Diversified industrial city

In China, a city in which a variety of industries have developed.

Do-it-yourself building materials

See HOME IMPROVEMENT.

Domestic loan

One of the principal sources of funds for construction in China, consisting of investment capital from domestic sources, such as the People's Construction Bank of China. See also PEOPLE'S CONSTRUCTION BANK OF CHINA and INVESTMENT IN CAPITAL CONSTRUCTION.

Domicile register

In China, the official record book at the local police substation in an urban area or kept by the township government in a rural area. The register contains the name, sex, dates of birth and death, moves, occupation, and kinship of the residents in each household within the given jurisdiction. See also POLICE SUBSTATION, REGISTERED RESIDENT,

区房管局 Qū fángguǎnjú

[见"市房管局"]

综合性工业城市 Zōnghéxìng gōngyè chéngshì

在中国,指以多种门类的工业企业为主的城市。

"自己动手"建筑材料 "Zìjǐ dòngshǒu" jiànzhù cáiliào

[见"住房改善"]

国内贷款 Guónèi dàikuǎn

中国建设项目资金的主要来源之一,利用国内资金(包括中国人民建设银行、中国人民银行及其他财政来源),对建设项目进行的投资。[参见"中国人民建设银行"、"基本建设投资"]

户籍 Hùjí

在中国,记载每户居民的姓名、性别、生死年月、迁移、职业、亲属关系等情况的册籍,在城市中,由有关派出所保存;在农村中,由乡政府保存。[参见"派出所"、"户口"、"户口簿"、"户口登记"、"转户"]

RESIDENCE CERTIFICATE, RESIDENT REGISTRATION, and TRANSFER OF DOMICILE REGISTRATION.

Dormitory suburb

See BEDROOM SUBURB.

Downpayment

In the United States, the percentage of the total sales price of a building or land to be paid in cash at the time of purchase.

Downtown

Term used, loosely in American everyday speech to designate the central area of business activity in a city or town. The actual area meant is not clearly defined and varies with context, especially in spread-out cities of the western United States, which may or may not have a recognized "downtown" area.

Downzoning

In the United States, zoning action that reduces the permissible intensity of land use or density of housing in order to control growth, for example, by increasing the minimum lot size. See also ZONING and ZONING AMENDMENT.

卧城区 Wòchéngqū

[参见"卧城郊区"]

首次付款 Shǒucì fùkuǎn

在美国,指购买房地产时,按其总售价的百分比于成交时先以现金偿付。

闹市区 Nàoshìqū

美国日常用语中泛指城镇中心的商业活动区,但其实际所指地区并未明确划定,可因情况而异。尤其在美西部分散式的城市中,有时并无公认的闹市区。

降低密度区划 Jiàngdī mìdù qūhuà

在美国,为了控制城市增长而采取的一种区划措施,即降低所允许的土地使用程度或住房密度。例如增加单块基地的最小面积。[参见"区划"、"区划修正"]

Drawdown

In the United States, the most common method of paying out money for construction financing. The lender establishes a line of credit for the borrower up to a maximum amount. The borrower then draws only the amount of money required for completion of a predefined stage of the project, paying interest only on the amount drawn to that point. See also CONSTRUCTION FINANCING.

Duplex apartment

An apartment with rooms on two floors and a private interior stairway.

Duplex house

A house that is separated by a common wall into two dwellings, each with its own facilities and entrance. Also called semidetached house. See also ATTACHED HOUSE.

Dwelling size

The size of a living unit, frequently described in terms of the number of habitable rooms per dwelling, the number of persons that the dwelling is designed to accommodate (e.g., two-person dwelling), or the number of bedrooms (e.g., three-bedroom house). Dwelling size may also be expressed in terms of floor area. See also AVERAGE FLOOR AREA PER

分期提款 Fēnqī tíkuǎn

在美国，指偿付建筑工程款项的一种最普通的做法。先由贷款人确定给予借款人的最高贷款额度。然后借款人按完成预定工程的阶段提取所需借款，并按当时所提金额支付利息。[参见"筹措建造资金"]

跃层公寓 Yuècéng gōngyù

每户为两层并有独用内部楼梯的公寓。

二联式住宅 Èrliánshì zhùzhái

一栋用隔墙分成两套且各自具有完善设备和入口的住房。亦称半独立式住宅。[参见"毗连住宅"]

户型 Hùxíng

指住户的住所类型，一般以每套住所的大小、所含居室数表示。有按可住人数表示的，如二人户；有按卧室数表示的，如三室户；亦有用建筑面积表示的。[参见"平均每套建筑面积"、"居住建筑面积"、"户室比"]

UNIT, RESIDENTIAL FLOOR AREA and SIZE DISTRIBUTION of DWELLINGS.

Dwelling unit

A separate living space, designed for residential occupancy of a family or individual, located within a multiunit structure or constructed as a separate building.

居住单元 Jūzhù dānyuán

为一个家庭或个人居住用的单独居住空间,或在多单元建筑物内,或建成单幢建筑物。

E

"Earnest money"

In the United States, cash payment in advance for part of the purchase price of property. Earnest money serves as a binder, i.e., proof of the purchaser's intent to enter into a contract.

Earth-sheltered housing

Residential structures that are constructed partially or totally underground or into hillsides with skylights and atriums for light and ventilation. The earth surrounding the dwelling is arranged so as to moderate temperatures, reduce noise from outside, and protect against diverse weather conditions such as tornadoes. See also CAVE DWELLING.

Easement

In the system of common law, the acquired right of one landowner to the use or enjoyment of another landowner's land. The use is specific and involves benefit without profit to the landowner, such as laying a sewer, putting up electric power lines, or crossing adjacent property to gain access. See also ACCESS RIGHT and RIGHT-OF-WAY.

Ecology

The relationships of all living organisms to each other and to their environment. Fundamental to

定金 Dìngjīn

在美国,指购买房地产时,按售价的一部分预付的现金。定金起保证作用,证明买主愿意签订合同。

掩土住房 Yǎntǔ zhùfáng

部分或全部建在地下或山坡内的居住建筑。设有天窗和内庭,用于采光和通风,并利用住宅四周的土层来调节温度,降低外部噪声和防御多变的气候条件,如龙卷风等。[参见"窑洞住宅"]

地役权 Dìyìquán

在习惯法中,指一地产主获得使用另一地产主的土地的权利,这种权利仅限于某些不妨碍地产主经济利益的方面,例如铺设下水道、架设电线或穿过相邻地产以便出入等。[参见"出入权"、"通行权"]

生态学 Shēngtàixué

研究生物之间和生物与其生存环境之间相互关系的一门科学。生态研究的基本概念在于生态系统,

ecological studies is the concept of an ecosystem, which may be interpreted as a biome (i.e., the whole complex of organisms, both animals and plants, naturally living together) and its habitat, e.g., a tropical forest ecosystem. All the interacting factors in a mature ecosystem are in appropriate equilibrium; it is through the interactions that the whole system is maintained. See also HABITAT, HUMAN ECOLOGY, and URBAN ECOLOGY.

Economic base

The sum of all activities that generate incomes in a defined planning area. According to economic base theory, the productive efforts of an urban area fall into two major categories: basic or primary activities that produce goods and services for export from the immediate area, thus bringing in new money, and service or secondary activities that simply recirculate financial resources already present. Basic industry is viewed as the key to economic growth. Economic base analysis quantifies and classifies economic activities in an area as a prerequisite to land use planning.

Economic obsolescence

See OBSOLESCENCE.

也可称为生物,即动物和植物自然并存的状态及其生存环境。例如热带丛林的生态系统,在成熟的生态系统中,所有相互作用的因素都处于适宜的平衡状态,通过这种平衡的相互作用,生态系统才得以维持。[参见"生长地"、"住地"、"人类生态学"、"城市生态学"]

基本经济　Jīběn jīngjì

指一个规划区内所有经济收入活动的总和,城市地区生产活动可分为两类。①基本的生产活动,即为本区生产对外销售的产品或为之提供服务的活动,从而得到新的资金。基本工业是一种积极的带来本区经济繁荣的关键因素。②服务性的收入活动。这是一个规划区内居于第二位的经济活动,仅起现有经济资源的再循环作用。基本经济的分析是对一个地区内的经济活动进行定量和分类,是制订土地使用规划的先决条件。

经济性淘汰　Jīngjìxìng táotài

[见"过时"]

Economic region

In China, a geographic area where industry and agriculture are closely linked and should be considered comprehensively in planning and development. The economic region consists of large, medium-sized, and small cities, rural areas, and possibly mine areas or giant projects, e.g., hydroelectric power stations. Within the region, certain cities become central because of their status and functions. See also CENTRAL CITY.

"Edible landscaping"

Use of ornamental herbs, vegetables, and fruit trees in plantings around residences to provide greenery that is attractive, functional, and nourishing.

EER

See ENERGY EFFICIENCY RATIO.

Effective demand

See HOUSING DEMAND.

Ekistics

As defined by C.A. Doxiadis, the science of human settlements, devoted to study of all types of human settlements from the isolated and nomadic hut to the organized farmstead, village, small town, city,

经济区 Jīngjìqū

在中国,工、农业经济发展上联系比较紧密,需要综合规划与开发的地区。其中也包括大、中、小城市及农村地区,有的可能包括矿区或巨型企业(如水电站等),有的城市因其地位和作用而成为该经济区的中心城市。[参见"中心城市"]

食用园林布置 Shíyòng yuánlín bùzhì

将可供观赏的药草、蔬菜、果树种植在住宅的四周,以提供悦目而有实用和营养价值的绿化。

能源效率比(缩写) Néngyuán xiàolǜbǐ(suōxiě)

[见"能源效率比"]

实际需求 Shíjì xūqiú

[见"住房需求"]

人类群居学 Rénlèi qúnjūxué

按 C.A. 多西亚蒂的定义,人类群居学是一门研究人类定居的科学。其具体内容是研究人类从独居和游牧者的小棚屋,到有组织的农舍、村庄、小城镇、城市、大都市、大都市群以及未来的城市

metropolis, megalopolis, and future major urban systems.

Elderly housing

In the United States, dwelling units specially designed and constructed to fill the specific needs of elderly residents. See also CONGREGATE HOUSING.

Emergency housing

Housing to accommodate homeless families or refugees after major disasters, such as those caused by floods, earthquakes, and war. In the United States, mobile homes are frequently used to house disaster victims. See also DISASTER PLANNING and SHELTER.

Eminent domain

In the United States, the right of the government to acquire property by condemnation for public or quasipublic use, provided that the owner is fairly compensated. The fifth amendment to the U.S. Constitution guarantees that the government cannot take property without just compensation to the owners. See also COMPENSATION FOR LAND, CONDEMNATION, and LAND ACQUISITION.

体系等的历史进程和各种定居方式。

老年住房 Lǎonián zhùfáng

在美国,指为适应年老居民的特殊要求而特别设计和建造的居住单元。[参见"集合公寓"]

应急住房 Yìngjí zhùfáng

为洪水、地震、战争等大型灾害造成无家可归的灾民所提供的一种住房。在美国,往往向灾民提供活动房屋作为应急住房。[参见"城市防灾规划"、"栖身处"]

国家征地权 Guójiā zhēngdìquán

在美国,指政府在付予公正的补偿条件下,为公用或半公用的需要而征用地产的权利。《美国宪法》第五条修正条款规定,政府若不对地产主付予公正补偿,则不得征用其产业。[参见"土地补偿费"、"征用"、"征地"]

Employee dormitory

In China, an institutionally owned residential building provided as living quarters for single employees who may be unmarried or married but not living together with their families. See also ENTITLEMENT TO HOUSING, FAMILY HOUSING, and OBLIGATION OF ORGANIZATIONS TO PROVIDE HOUSING.

Employment generation

In the United States, creation of jobs, usually through specific public or private investment decisions or programs which result in revitalization of depressed areas with low income levels and high unemployment rates.

Encroachment

In the United States, the unauthorized expansion of a structure onto land that does not belong to the owner of the structure.

Energy conservation

The exercise of public or private, voluntary or obligatory measures to curb consumption of energy resources. In the United States, such measures include lower highway speed limits, more fuel-efficient automobiles, lower thermostat settings in buildings, and increased structural insulation. See

单身宿舍 Dānshēn sùshè

在中国，指单位供其单身职工居住的住房。单身职工指未婚或已婚而不与家属同住在一处者。[参见"住房申请权"、"家属宿舍"、"住房分配"]

提供就业 Tígōng jiùyè

在美国，指通过公家或私人投资计划创造的就业机会，从而使低收入和高失业率的不景气地区复苏。

侵占 Qīnzhàn

在美国，指建筑物的业主未经许可而将自己的建筑物扩展到他人的土地上。

能源保护 Néngyuán bǎohù

公家或私人，自愿地或被迫地采取控制能源消耗的各种措施。在美国，这种措施包括降低公路行车速度、采用节油汽车、降低建筑物内恒温装置的标准以及提高房屋的保温隔热性能等。[参见"保护"]

also CONSERVATION.

Energy efficiency ratio (EER)

A value that represents the electrical efficiency of air conditioners and appliances, obtained by dividing Btu-per-hour output by electrical watts input during cooling.

Energy performance standard

In the United States, a criterion for optimum efficiency of energy use in buildings, usually established by assessing how much energy certain types of buildings are currently designed to use and how much energy they can be designed to use. See also PERFORMANCE CODE.

Engineer's approval

See ARCHITECT'S APPROVAL.

Enterprise zone

In the United States, a distressed area designated by the government as a special low-tax zone at least partially exempt from some regulatory constraints in order to create a climate of opportunity conducive to new business activity. The enterprise zone was introduced in recent times on the international level in Hong Kong, then adapted on a local level in Great

能源效率比　Néngyuán xiàolǜbǐ

在美国，指表示空气调节器和设备的电效率的数值，即降温时每小时输出的热量（BTU）除以每小时输入的电量（瓦）。

能量效率标准　Néngliàng xiàolǜ biāozhǔn

在美国，指建筑物中能量使用达到最佳效率的标准，通常是对几种类型建筑物的现行设计中所需的能耗以及可以达到的能耗做出评定后取得。［参见"功能法规"］

工程师认可证明　Gōngchéngshī rènkě zhèngmíng

［见"建筑师认可证明"］

兴业区　Xīngyèqū

在美国，指由政府定为"低税区"的经济不景气地区，可部分地免除某些规定的限制，以创造机会促进新的商业活动。兴业区已在香港被采用，其活动范围为国际性的，继而在英国被采用，但仅为地方性的。在美国，这一方法已由联邦政府建议，并被26个州采纳，作为在衰退的市区内邻里中增加就业机会和促进投资的一种手段。

Britain. In the United States, this approach has been proposed at the federal level and adopted in twenty-six states as a means of stimulating jobs, investment, and entrepreneurship in deteriorating inner city neighborhoods.

Entitlement city

In the United States, an urban area ruled eligible to receive funds or grants for projects under programs such as the Community Development Block Grant program. See also BLOCK GRANT and DISTRESSED CITY.

Entitlement to housing

In China, the right of all registered urban residents to apply for and to be assigned to housing, either by the institutions where they work or by the local housing management authorities when individuals are not in a position to apply to any institution. See also EMPLOYEE DORMITORY, FAMILY HOUSING, MUNICIPALLY OWNED HOUSING, MUNICIPAL HOUSING MANAGEMENT BUREAU, OBLIGATION OF ORGANIZATIONS TO PROVIDE HOUSING, and RIGHT OF OCCUPANCY.

Environmental design

Planning of land use that is based on fundamental

受资助城市 Shòuzīzhù chéngshì

在美国,指由政府规定有资格接受计划项目资金或赠款的城市地区,例如社区开发一揽子赠款计划等。[参见"一揽子赠款"、"贫困城市"]

住房申请权 Zhùfáng shēnqǐngquán

在中国,城市中有正式户口的居民所享有的申请住房的权利。职工可向其工作单位申请;不能向工作单位申请者,可向城市房管部门申请。[参见"单身宿舍"、"家属宿舍"、"房管部门住房"、"市房管局"、"住房分配"、"居住权"]

环境设计 Huánjìng shèjì

以物质环境的质量为考虑的基本点,以优良环境

concern with the quality of the physical environment and on the premise that a good environment is a basic human right and necessity. Professions devoted to environmental design activities are architecture, engineering, landscape architecture, and urban planning. See also ARCHITECTURE, CARRYING CAPACITY, DESIGN WITH NATURE, LANDSCAPE ARCHIERCTURE, and URBAN DESIGN.

Environmental impact statement

In the United States, a study of the environmental consequences of a project or policy, as compared to alternative projects or policies. Environmental impact statements must by law be prepared by federal agencies for major projects such as large-scale, multifamily subdivisions to facilitate decisions on their acceptability from an environmental viewpoint. See also ENVIRONMENTAL STANDARDS and IMPACT.

Environmental standards

Legislation, regulations, or guidelines controlling use and/or protection of natural resources and values. Examples are laws and regulations establishing acceptable air and water quality levels to protect human health or establishing and preserving park land or open space. See also ENVIRONMENTAL

是人的基本权利与需要为前提的土地利用规划，环境设计工作所包含的专业有建筑学、土木工程学、风景建筑学及城市规划。[参见"建筑学"、"持续能力"、"适应自然的设计"、"风景建筑学"、"城市设计"]

环境影响报告 Huánjìng yǐngxiǎng bàogào

在美国，指通过多方案或不同政策的比较，对建设项目所产生的环境效果进行研究的报告。重点项目的环境影响报告，按法律规定，必须由联邦政府机构提出，例如从环境观点出发，对把土地大规模地再划分给多户是否可取进行分析，以便做出决定。[参见"环境标准"、"影响"]

环境标准 Huánjìng biāozhǔn

为了控制自然资源及财富的使用与/或保护而制定的法律、规范或准则，例如为保护人类健康而制定的空气与水的质量标准的法规与规范，或保存园林绿地的法规与规范。[参见"环境影响报告"、"功能法规"、"功能标准区划"]

IMPACT STATEMENT, PERFORMANCE CODE, and PERFORMANCE STANDARD ZONING.

Equity

1. The excess of a company's assets over its liabilities.

2. In the United States, the current market value of property minus the unpaid mortgage balance and any other claims. See also SWEAT EQUITY.

3. A share in the issued capital of a company. Shareholders investing in the company are entitled to vote at annual meetings and to share in profits through dividends but are also liable for business losses. Also called equity capital and shareholder's equity.

Escalator clause

In the United States, a contractual provision for the upward or downward adjustment of wages, rent, construction costs, or other contractual payments based on a specific price index. For example, a rental agreement may stipulate that payments will be increased as property taxes or operating expenses rise.

Escrow

In the United States, money, property, or securities held by a disinterested third party to ensure

资产净值 Zīchǎn jìngzhí

1. 一个公司的资产减去其负债后的净额。
2. 在美国,指房地产的市面价值扣除未偿还的抵押款和任何其他债务后的净额。[参见"改进后净增值"]
3. 指一个公司所发行股票的净值。凡该公司股票持有者(投资者),在股东年会中,享有投票权和分红权,同时也负有营业损失的责任。资产净值又称"自有股本"、"股东自有股本"。

调整条款 Tiáozhěng tiáokuǎn

在美国,指合同中的一种条款,其中规定根据特定物价指数对工资、租金、建筑成本或其他合同付款作必要增减调整,例如租约规定租金可随财产税或经营费的上涨而增加。

寄存担保品 Jìcún dānbǎopǐn

在美国,指由订立合同的一方将现金、房地产或证券等寄存在与双方无利害关系的第三者处,作

fulfillment of contractual obligations of one party or organization to another. Funds deposited to this end in a special account are said to be placed in escrow.

Expediter

In the construction industry, an individual or organization employed to coordinate the procurement and delivery of materials and equipment required for construction. The purpose of the position is to avert material or equipment shortages that might result in delays in the schedule or costly errors in sequencing of work elements.

Expressway

In the United States, a high-speed, multilane divided highway for long-distance through-traffic. Access to the expressway from other roads is semicontrolled through the use of cloverleaf ramps, and cross-traffic is reduced by the use of overpasses and underpasses at intersections. See also BELTWAY, BYPASS, FREEWAY, GRADE SEPARATION, INTERSTATE HIGHWAY SYSTEM, and PARKWAY.

Extended family

A family in which more than two generations of directly related family or relatives are living together. The extended family used to be the most

为对另一方履行合同义务的保证品。为此目的而存储的资金专立账户,称为"寄存在第三者处的保证金"。

器材调度员 Qìcái diàodùyuán

在建筑业中,被雇用来协调采办和运送工程所需的材料和工具的人或组织,以避免因材料和工具的短缺而延误工程进度,或因施工程序安排不妥而导致代价昂贵的失误。

高速公路 Gāosù gōnglù

在美国,指长距离直达运输的高速、多车道公路。采用苜蓿叶式的匝道以控制自其他道路进入高速公路,并在道路交叉口处采用上跨和下穿式道路以减少平交。[参见"环城公路"、"绕行路"、"快速干道"、"立体交叉"、"州际公路网"、"林荫公路"]

大家庭 Dàjiātíng

由居住在一起的两代或两代以上直系家属或其他亲属组成的家庭。在中国,大家庭曾是最普遍的一种家庭形式。[参见"家庭"、"核心家庭"]

common family type in China. See also FAMILY and NUCLEAR FAMILY.

Extension services

In the United States, educational opportunities, information, and technical training provided to the general public through special programs or projects, sponsored by an institution such as a university and made readily available through facilities within the community.

Externalities

The indirect consequences, either positive or negative, of a course of action, affecting groups or sectors other than the intended beneficiaries of that course of action. Analysis of real or potential side effects may prompt alteration of private or public decisions or programs, i.e., the decision to relocate a company or a program to construct a highway through a particular part of a city. See also FORECASTING, IMPACT, SIDE EFFECT, and RISK ANALYSIS.

Extra expenses

In China, a constituent part of the construction cost together with direct and indirect expenses. Extra expenses include such items as costs for temporary

附设服务 Fùshè fúwù

在美国,指根据特别计划或项目,由大学等机构主办,通过设在社区内的设施向居民提供的服务,例如学习机会、信息提供和技术培训等。

外在副作用 Wàizài fùzuòyòng

行动过程中,对于该行动的预期受益人以外的集团或部门所产生的间接的、正面的或反面的效果。对于这些实际的或潜在的副作用的分析,可能导致私人或公家的某些决策或计划发生变更。例如,由于决定在城市的某一地区建造一条公路而必须搬迁一家公司或一项工程。[参见"预测"、"影响"、"副作用"、"风险分析"]

独立费 Dúlìfèi

在中国,指工程成本的组成部分。即工程成本中除直接费和间接费以外的部分,包括临时设施费、流动施工津贴费、副食补贴费、劳保支出等。[参

construction facilities; subsidies to staff and workers for relocation and nonstaple food extras, and insurance for workers. See also CONSTRUCTION COST, DIRECT EXPENSES, and INDIRECT EXPENSES.

Exurbia

In the United States, the sparsely settled, semirural unclearly demarcated region beyond the suburbs but within the zone of influence of a city. The population of the region may include exurbanites, a term coined by A.C. Sepectorsky for wealthy persons who live in exurbia and work in the city. However, the composition of the population is generally varied, including farmer and residents of older rural homes. The meaning of the term thus varies widely from speaker to speaker.

见 "建筑成本"、"直接费"、"间接费"]

城市远郊　Chéngshì yuǎnjiāo

在美国,指城市郊区以外,但仍处于城市影响范围以内,人口稀少,无明确界限的半农村地区。在该地区内的人口组成中,既包括住在远郊而在城里工作的富有者(即 A.C. 斯派克托尔斯基称之为"远郊的富有居民"),也包括农民和当地的老住户。该词语义常因使用者而异。

F

Fair housing

A policy, implemented in the United States by law, that prohibits discrimination in the purchase, sale, or ownership of real estate on the basis of race, color, religion, sex, national characteristics, or demographic characteristics such as age.

Fair market value

See MARKET PRICE.

Family

For purposes of computing densities and urban census analysis, two or more persons related by blood, birth, marriage, or adoption who occupy the same dwelling. In general usage, the family includes related persons regardless of where they live. In China, family usually refers to directly related family in the same line, i.e., grandparents, parents, and children. The Chinese Constitution as well as the Marriage Law and Inheritance Law of China all stipulate that children have a legal obligation to support their parents. In the United States, the most commonly occurring family unit consists of two adults with or without children, but may include other relatives or be headed by only one adult. The U.S. Housing and Community Development Act of 1974 also defines as a family two or more unrelated, elderly, disabled,

公平住房政策 Gōngpíng zhùfáng zhèngcè

在美国，根据法律执行的一项政策，规定房地产的买卖或所有权不得因种族、肤色、宗教、性别、国籍或人口统计上的特征（如年龄）而有任何歧视行为。

公平市场价值 Gōngpíng shìchǎng jiàzhí

［见"市场价格"］

家庭 Jiātíng

在统计城市人口密度和人口普查分析时，指因血统、出生、婚姻或领养关系而居住在同一住处的两人或多人。在一般用语中，指有亲属关系而不论住处的人们。在中国，家庭通常指直系血亲，如祖父母、父母、子女等。按《中国宪法》、《婚姻法》和《继承法》规定，子女有赡养其父母的法定义务。在美国，最普遍的家庭是两个成人，无论有无子女，但也可包括其他亲属，或只有一个成年户主。按美国1974年的《住房和社区发展法》，也可指住在一起而无血亲关系的两个或几个老年、生活不能自理者或残疾者。［参见"大家庭"、"户"、"核心家庭"、"独生子女户"］

or handicapped individuals living together. See also EXTENDED FAMILY, HOUSEHOLD, NUCLEAR FAMILY, and ONLY-CHILD FAMILY.

Family housing

In China, housing built by organizations, institutions, and enterprises for cadres, workers, and their families, as opposed to dormitories for single employees. See also EMPLOYEE DORMITORY, ENTITLEMENT TO HOUSING, and OBLIGATION OF INSTITUTIONS TO PROVIDE HOUSING.

Family land plot

In China, a small piece of land allotted to the family of a cooperative member for its own long-term use after land reform by the former agricultural producers' cooperative.

FAR

See FLOOR-AREA RATIO.

Fast track construction

Starting construction of a building before all design details are complete. The key to the success of this building method is the design procedures followed by structural engineers. All buildings of modern design in the United States today are erected using this cost-

家属宿舍 Jiāshǔ sùshè

中国的机关、事业、企业单位为其职工带眷居住所建造的住房（有别于单身宿舍）。[参见"单身宿舍"、"住房申请权"、"住房分配"]

自留地 Zìliúdì

在中国，由过去的农业生产合作社在土地统一经营以后，分配给社员家庭长期使用的少量土地。

建筑面积比（缩写） Jiànzhù miànjībǐ (suōxiě)

[见"建筑面积比"]

快速施工法 Kuàisù shīgōngfǎ

一种在设计详图未全部完成之前即开始施工的方法。其成功关键在于结构工程师所采取的设计步骤。目前美国设计的现代建筑均采用这种节约费用的施工方法。

saving technique.

Feasibility study

A detailed investigation and analysis conducted to determine the financial, economic, or technical advisability of putting a particular plan or project into effect.

Fee

Charge to an individual or organization for a particular service or privilege. A professional fee represents payment for professional ability, capability, and availability of an organization or individual. In construction work, the charge for professional services may be a percentage fee, i.e., a defined percentage of the construction costs. Fees may also be charged for services such as check processing (service fees) or for special licenses (licensing fees). See also PERCENTAGE AGREEMENT and USER FEE.

Feedback

The return of output from a system, machine, or process to the input level to evaluate whether the desired results have been achieved. Feedback may also designate explicit responses or reactions to a project or event that are channeled back to the

可行性研究 Kěxíngxìng yánjiū

对于拟建工程的详细调查和分析,以便确定该工程在财务、经济、技术或其他方面的可取性。

费 Fèi

为提供某种服务或权益而向个人或团体收取的费用。例如,专业服务费是向团体或个人提供专业性的技能、才智和效能所收取的报酬。在建筑工程中,专业服务费常按建筑成本的一定百分数计取;此外,还包括各种服务的手续费,如处理支票的服务费或申请特殊执照的执照费等。[参见"按百分比收费协议"、"用户费"]

反馈 Fǎnkuì

使一个系统、机器或过程的输出返回到输入的水平上,以评价其是否达到预期目的。反馈也可以指返回到原工程或事件发起人处的明确回答反应。

initiators.

Feeder bus system

In the United States, a bus transportation system with a number of local bus lines that feed into terminals connecting with main lines.

Field order

In the United States, a written order effecting a minor change in the work that does not involve an adjustment to the contract sum or an extension of the contract time, issued by the architect-engineer to the contractor during the construction phase.

Field survey

Collection of data of a particular type, for example, on land use, at the sites or locations under study.

Final acceptance

Inspection and approval of a construction project or item after its completion. In the United States, the owner accpets the project from the contractor after the architect or engineer certifies that it is complete and meets contract requirements. In China, the project is usually approved by a group of architects, engineers, and personnel of other specialties. See also ACCEPTANCE OF WORK SUBELEMENTS,

公共汽车支线系统 Gōnggòng qìchē zhīxiàn xìtǒng

在美国,指由一些地方性公共汽车线路构成的运输系统,通过终点站与主线相连。

变更通知 Biàngēng tōngzhī

在美国,指施工阶段由建筑师或工程师签发给承包商的关于工程微小变更的书面通知。这种变更不影响合同总金额或工期。

现场调查 Xiànchǎng diàochá

指在现场搜集某种数据的工作。例如,在所调查的现场或地点搜集有关土地利用方面的数据。

最终验收 Zuìzhōng yànshōu

对已完成工程所进行的检查验收。在美国,业主根据建筑师或工程师所开具说明工程完工并符合合同要求的证明书,即对工程验收。在中国,此项工作一般由业主及建筑师、工程师及其他专业人员一起进行。[参见"分项工程验收"、"隐蔽工程验收"、"全部竣工项目"、"建筑缺陷清单"]

ACCEPTANCE OF HIDDEN SUBSURFACE WORK, COMPLETED PROJECT, and PUNCH LIST.

Finance

Capital in the form of funds lent or borrowed, usually for investment, through financial markets or institutions.

Financial planning

Establishment of a plan for effective management and investment of the funds of an individual or organization.

Fixed-asset investment

In China, a general term for investment in capital construction and industrial modernization, consisting of cost of construction work, expenses for procurement of equipment and tools, and miscellaneous expenses. See also INDUSTRIAL MODERNIZATION INVESTMENT and INVESTMENT IN CAPITAL CONSTRUCTION.

Fixed-price competition

In the United States, a method used for contracting out work. The price of the project is set in advance, and selection of the contractor is based on other

筹资 Chóuzī

为了投资而通过金融市场或机构以借贷形式筹措到的资金。

制订财务计划 Zhìdìng cáiwù jìhuà

为了使个人或团体的资金得到有效的管理和投资而制订计划。

固定资产投资 Gùdìng zīchǎn tóuzī

中国基本建设投资和更新改造措施投资的总称,内容有：①建筑安装工程费用；②设备、工具、器具购置费用；③其他投资费用。[参见"更新改造措施投资"、"基本建设投资"]

固定价格竞争 Gùdìng jiàgé jìngzhēng

在美国,指发包工程的一种方法,即先确定工程造价,而对承包人的选择则根据其他因素考虑,如设计方案的优劣、承包公司的资格以及实施方

factors, e.g., design, company qualifications, and specific benefits of a particular approach. The federal government uses this technique to offer urban land for redevelopment.

Fixed-price contract

A contractual agreement in which the amount of money to be paid for goods, services, or some type of transaction is set and not subject to subsequent change.

Fixed-rate mortgage

In the United States, a form of longterm loan to the home buyer, with a fixed term, fixed interest rate, and fixed monthly payments according to an amortization schedule, usually for repayment within thirty years. The fixed-rate mortgage was previously called a standard mortgage, as it was the usual form of home financing. See also AMORTIZATION and MORTGAGE.

Flat

See APARTMENT.

Flatted factory

See MULTISTORY FACTORY.

式所带来的利益等。联邦政府为城市改造提供土地时常采用这种方法。

固定价格合同 Gùdìng jiàgé hétóng

合同的一种方式。合同规定对于货物、服务或某种交易所付的金额是固定的,事后不得更改。

固定利率抵押贷款 Gùdìnglìlǜ dǐyā dàikuǎn

在美国,指发放给购房者的一种长期贷款。以固定的贷款期限、贷款利率及每月偿还金额、按分期偿还时间表偿还本息(通常在30年内)。是筹措购房资金的常用方法,故以前称标准抵押贷款。[参见"分期偿还"、"抵押贷款"]

套房 Tàofáng

[见"公寓"]

多层出租厂房 Duōcéng chūzū chǎngfáng

[见"分层出租厂房"]

Floating zone

In the United States, a land area established on the basis of performance standards listed in zoning ordinances but not shown on the zoning map until the area has been rezoned for development.

Floodplain

A flat land area which may be submerged during floods and must therefore receive special attention in development planning.

Floor area

See GROSS FLOOR AREA.

Floor-area ratio(FAR)

In the United States, an index expressing the total permitted floor area of a building as a multiple of the area of the lot, used for regulating building bulk. For example, a FAR of 2 for a lot 15,000 square feet in area would permit a building with 30,000 square feet total area. Comparable indices in Great Britain are the plot ratio and the floor-space index. See also BULK REGULATIONS.

Floor-space index(FSI)

See FLOOR-AREA RATIO.

浮动区 Fúdòngqū

在美国,指根据区划法令中所列的功能标准而划定的一块土地,但在为正式开发而进行重新区划前,这块土地在区划图上不予标明。

泛滥平原 Fànlàn píngyuán

可能遭受洪水淹没的平坦地区,在开发规划中应予以特别注意。

建筑面积 Jiànzhù miànjī

[见"建筑毛面积"]

建筑面积比 Jiànzhù miànjībǐ

在美国,指控制建筑物体积的指标,将允许的总建筑面积以用地面积的倍数表示。例如,在15000平方英尺的土地上,如建筑面积比为2,即表示其上所建的总建筑面积不得超过30000平方英尺。在英国,其相应指标称为"容积率"和"建筑面积指标"。[参见"体积规定"]

建筑面积指标 Jiànzhù miànjī zhǐbiāo

[见"建筑面积比"]

Force account

1. In the United States, the part of an expense account of a local government that results from direct employment of special-purpose labor forces for such public services as garbage collection and road maintenance instead of contracting out such services to commercial firms.

2. In the United States, work performed under pressing circumstances without a prior cost agreement and billed at the cost of labor, equipment, materials, and other expenses plus an agreed percentage for profit and overhead.

Forecasting

The attempt to predict future events or conditions through rational analysis of pertinent data and such techniques as economic modeling. See also EXTERNALITIES and RISK ANALYSIS.

Foreclosure

In the United States, legal action canceling a mortgagor's right to redeem a mortgage. Action is usually brought by a lender or lending institution after the borrower has defaulted in payment under a mortgage agreement. If the court finds for the lender, it issues a decree establishing the debt and arranging for public sale of the mortgaged property. See also

1. 自营项目开支 Zìyíng xiàngmù kāizhī

在美国,指地方政府开支费用中的一项,即对某些公共服务项目,如收集垃圾、保养道路等,不是发包给商业服务公司,而是直接雇用劳动力所需的开支。

2. 成本加费用账款 Chéngběn jiā fèiyòng zhàngkuǎn

在美国,指在紧迫情况下进行工程项目时,因无事先协议的价格可循,因而按人工、设备、材料及其他成本费用,外加经双方同意的一定百分比的利润和管理费用进行结算的账款。

预测 Yùcè

根据对有关数据的合理分析,以及通过例如经济模型等技术,对未来事件或情况做出估计。[参见"外在副作用"、"风险分析"]

取消赎回权 Qǔxiāo shúhuíquán

在美国,指取消抵押人赎回其抵押品的权利的法律行动。这种情况常因借款人不按抵押契约的规定付款,而由贷款人的个人或机构提出诉讼。如法院判决贷款人胜诉,则发出判决书,确认该项债务,并安排公开拍卖该项抵押产业。[参见"违约"、"拖欠抵押贷款"]

DEFAULT and DELINQUENT MORTGAGE.

Foreign loan

Investment capital from abroad, one of the financial resources for construction projects in China. Foreign investment capital is generally raised by the Bank of China and the China International Trust & Investment Corporation. See also INVESTMENT IN CAPITAL CONSTRUCTION.

Foreman

In the U.S. construction industry, an employee at a construction project who supervises a group of workers and usually is responsible to a superintendent or manager. A general foreman is the contractor's representative at the construction site who, as supervisor of the construction labor force, coordinates the foremen of various skill groups within the construction crew.

Formula funding

Allocation of grant money by the U.S. federal government using a statutory formula to define the recipients' needs on the basis of relevant demographic, social, and economic data. See also BLOCK GRANT, COMPETITIVE FUNDING, and DISCRETIONARY FUNDING.

国外贷款 Guówài dàikuǎn

来自国外的投资,是中国建设项目资金来源之一。中国一般通过中国银行、中国国际信托投资公司等筹集外资。[参见"基本建设投资"]

工长 Gōngzhǎng

在美国建筑业中,受雇在一项工程中领导一个队组工人,并对工段长或主任负责的人员。总工长是承包人驻工地的代表,总管全部工人,负责协调各专业工长的工作。

按公式拨款 Àn gōngshì bōkuǎn

由美国联邦政府按法定公式根据有关的人口、社会和经济方面的数据,核定受款者的需要,给予津贴。[参见"一揽子赠款"、"竞争性拨款"、"选择性拨款"]

Freestanding new town

A self-sufficient development away from existing cities which provides housing and community facilities for persons employed in the community or in resource-based industries such as mining or forestry. See also NEW TOWN.

Freeway

In the United States, a multilane divided highway with fully controlled access for intersecting roads. The terms "freeway" and "expressway" are often used interchangeably in everyday speech, although theoretically access control is more complete on a freeway than on an expressway. Called a motorway in Great Britain. See also EXPRESSWAY and GRADE SEPARATION.

Frontage

1. The part of land or property that faces a street.
2. The length of the boundary of land or property that parallels the street, measured in linear feet or meters, e.g., fifty-foot frontage.

Functional obsolescence

See OBSOLESCENCE.

独立新城　Dúlì xīnchéng

远离现有城市而自给自足的新建社区。专为在本社区或资源工业部门（如矿业或林业）就业的人员提供住房和社区设施。[参见"新城"]

快速干道　Kuàisù gàndào

在美国，指交叉路入口完全受到限制的多车道高速公路。在日常用语中，"快速干道"与"高速公路"同义，但在理论上，前者的入口控制较后者更为彻底。在英国称为机动车路。[参见"高速公路"、"立体交叉"]

临街面　Línjiēmiàn

1. 一块土地或房地产临街的一面。
2. 土地或房地产沿街的地界长度，以英尺或米度量，如50英尺长的临街面。

功能性过时　Gōngnéngxìng guòshí

[见"过时"]

Functional planning

Establishment of goals, policies, and procedures, especially by government, for specific areas of need or activity, for example, transportation, housing, and water quality.

功能规划 Gōngnéng guīhuà

对某些领域（如运输、住房和水质）的需要或活动定出目标、政策和工作程序的规划，通常由政府制订。

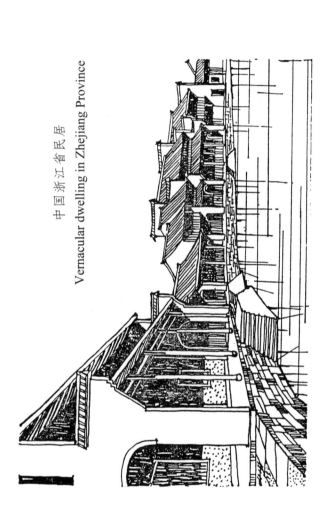

中国浙江省民居
Vernacular dwelling in Zhejiang Province

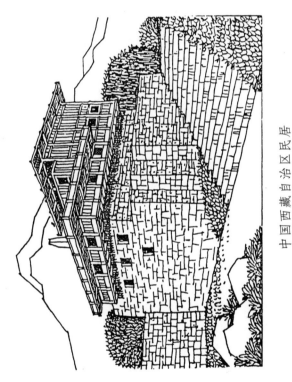

中国西藏自治区民居

Vernacular dwelling in Xizang Autonomous Region

G

"Gap financing"

Funds necessary to complete a project which are not available from the primary debt source or from equity investment. While these funds generally represent a small portion of the financing for a specific project, they are critical to its realization. Private sources of gap financing often require special guarantees or security arrangements. See also CONSTRUCTION FINANCING.

Garden apartment

In the United States, a living unit in one of a succession of buildings two-to-four stories high. The buildings are usually surrounded by open green areas and landscaping, and individual apartments may have balconies or verandas. See also APARTMENT.

General contract

A written agreement between the owner and a contractor who is to provide a service or product. In the construction industry, a general contract may be for construction of the entire building or be limited to architectural and/or structural work. See also CONTRACT, CONTRACTOR, and PRIME CONTRACT.

General contractor

See CONTRACTOR.

资金缺口 Zījīn quēkǒu

为完成某一项工程所必需而又不能从主要借款来源或股票投资中获得的资金。这种资金通常仅占一项工程所需总资金的一小部分,但对完成工程却起关键作用。向私人筹借该种资金往往需要特殊保证或提供担保品。[参见"筹措建造资金"]

花园公寓 Huāyuán gōngyù

在美国,指房屋周围有开阔的绿地和风景的二至四层的住宅楼。每套公寓可有阳台或廊子。[参见"公寓"]

总承包合同 Zǒngchéngbāo hétóng

业主与提供服务或产品的承包人之间所签订的协议。在建筑行业中,总承包合同可以指对全部工程的承包,也可以指对建筑或结构工程的承包。[参见"合同"、"承包人"、"直接承包合同"]

总承包人 Zǒngchéngbāo rén

[见"承包人"]

General foreman

See FOREMAN.

Gentrification

In Great Britain and the United States, the immigration of middleclass families into blighted neighborhoods in inner cities that were formerly dominated by poor or working-class families. The new residents use their own savings to restore houses, stimulating urban revitalization without a massive infusion of public funds but often resulting in displacement or relocation of poor families.

GFA

See GROSS FLOOR AREA.

Ghetto

In the United States, a neighborhood, usually in an inner city area, populated primarily by members of one or more minority ethnic or racial groups with common cultural values. Ghetto inhabitants tend to have low educational levels, poor job skills, and low incomes, and to remain separate from the mainstream of society. See also INNER CITY.

Government agency

A general term for administrative units on different

总工长 Zǒnggōngzhǎng

［见"工长"］

中产阶级化 Zhōngchǎnjiējíhuà

在美国和英国，指中产阶级家庭迁入到城内衰落地区中以往大多为贫民和工人阶级家庭居住的邻里中去。迁入的中产阶级居民以他们自己的积蓄修复住宅，因此不需投入大量公共资金即可刺激城市复苏，但这种做法往往导致贫苦家庭的迁居。

建筑毛面积（缩写） Jiànzhù máomiànjī (suōxiě)

［见"建筑毛面积"］

少数人种聚居区 Shǎoshùrénzhǒng jùjūqū

在美国，指一个或几个共同文化背景的少数民族聚居的邻里，通常位于城内衰落区。这种居民大多教育水平低、工作技艺差、收入微薄，并脱离社会的主流。［参见"城内衰落区"］

政府机关 Zhèngfǔ jīguān

各级政府机构的泛称，以有别于事业单位和企业。

level of the government, as opposed to institutes and enterprises.

GPM

See GRADUATED PAYMENT MORTGAGE.

Grade separation

In the United States, the use of underpasses and overpasses at junctions of a highway with other highways, pedestrian paths, or railroad tracks. Grade separations are usually constructed along the route of expressways and freeways to maintain a steady, rapid flow of traffic. See also EXPRESSWAY and FREEWAY.

Grading

1. Arrangement of people, organizations, or products into ascending or descending levels according to rank, quality, price, production, or some other criterion.
2. Modification of the ground surface, for example, of a roadbed or lot, to an even level or to a progressive ascent or descent by cut and/or fill work.

Graduated payment mortgage (GPM)

In the United States, a mortgage plan under which only a small downpayment is required and initial

递增偿还抵押贷款（缩写） Dìzēng chánghuán dǐyā dàikuǎn (suōxiě)

[见"递增偿还抵押贷款"]

立体交叉 Lìtǐ jiāochā

在美国，指在公路与公路、步行道或铁路线的交点上，沿高速公路和快速干道的走向设置上跨路和下穿路，使交通保持稳定快速。[参见"高速公路"、"快速干道"]

1. 分级 Fēnjí

根据地位、质量、价格、生产量或其他标准，对人口、机构或产品由低到高或由高到低地划分等级。

2. 平整土地 Píngzhěng tǔdì

整理地表，如通过填挖土方将路基或场地整成水平或上、下坡面。

递增偿还抵押贷款 Dìzēng chánghuán dǐyā dàikuǎn

在美国，指一种分期偿还的抵押贷款方式。其首次交付的金额很少，初期每月偿还的金额较低，

monthly loan payments are low, increasing over a specified period (e.g., five years) and then leveling off for the remainder of the thirty-year term of the loan. See also MORTGAGE.

Grant

Donation or allocation of money or property by an individual, organization, or government for a specific purpose such as research or development. See also BLOCK GRANT, MATCHING GRANT, and URBAN DEVELOPMENT ACTION GRANT.

Green space

See OPEN SPACE.

Greenbelt

A belt-like circle of open land, parks, or farm land preserved by official authority around a city or town to limit urban growth and merging of densely built-up areas.

Gridiron plan

1. In the United States, a land use planning map divided into sectors by uniformly spaced, intersecting horizontal and vertical lines. Also called a grid.
2. A regular pattern of street layout in which streets intersect each other at right angles.

在以后的一段时间（如五年）内偿还金额逐渐递增，然后再以每年同样的金额偿付完 30 年期限的贷款。[参见"抵押贷款"]

赠款 Zèngkuǎn

为了达到特定的目标，例如研究或开发，由个人、团体或政府捐赠或拨给的款项或产业。[参见"一揽子赠款"、"相应赠款"、"城市开发活动赠款"]

绿地 Lùdì

[见"空地"]

绿带 Lùdài

指围绕着城市，由园林或农田等组成的带状地区，一般由官方机构规定，用以限制城市的扩展，防止人口稠密的建成区之间连成一片。

1. 网格规划图 Wǎnggé guīhuàtú
在美国，指用横竖的交叉线将土地划分成相同地块的土地利用规划图，亦称网格。

2. 方格式路网 Fānggéshì lùwǎng
街道互成直角的规整布局。

Gross floor area (GFA)

In the United States, the most commonly used term for the total floor area of all stories of a building in square feet, measured from the outside of the exterior walls and excluding unenclosed areas. Net floor area is calculated from the inner faces of containing walls.

Gross leasable area

See LEASABLE AREA.

Gross residential density

See RESIDENTIAL DENSITY.

Ground lease

See LEASE.

Group house

In the United States, a living arrangement in which a group of unrelated individuals share rent and expenses for a house.

Growth center

Primary focus of growth, usually in an urban area in a developing country. The concept of a growth center is closely linked to the growth pole theory of F. Perrouy. In this view, a growth pole is a group of industries clustered around one or more dynamic

建筑毛面积 Jiànzhù máomiànjī

在美国,指建筑物各层建筑面积之和的常用词,以平方英尺计,自外墙的外表面算起,不包括无外墙封闭的面积。建筑净面积则从外墙的内表面算起。

可出租毛面积 Kěchūzū máomiànjī

[见"可出租面积"]

居住毛密度 Jūzhù máomìdù

[见"居住密度"]

土地租约 Tǔdì zūyuē

[见"租约"]

合租住宅 Hézū zhùzhái

在美国,指由相互无关系的一些个人共同租用并分摊租金和费用的住房。

增长中心 Zēngzhǎng zhōngxīn

通常指发展中国家城市地区内经济增长的主要集中点。这一概念与F.佩罗的"增长极"理论有密切关系。所谓增长极是指围绕着一个或几个充满生机、能使该地区内其他地方产生经济活力的工业而形成的一组工业。增长中心也是一个吸引中

industries which generate economic vitality to the rest of the region. The growth center also serves as an attraction pole, with the power to attract resources, equipment, or means of economic exchange within a given zone of influence.

Growth rate

Statistical estimate of the increase in size or population of an entity or area in a given time period.

Guaranty

1. In the United States, a legally binding written assurance by a manufacturer that a product meets certain quality standards and will be replaced if it proves defective within a specified period. Also called a warranty. See also HOME WARRANTY.

2. In the United States, a written promise by a person or organization to bear the responsibility for the debt or obligation of another person or organization in case of default or nonperformance.

心，在一定的影响范围内具有吸引资源、设备或经济交流的能力。

增长率 Zēngzhǎnglǜ
对某一实体或地区的规模或人口在一定时期内的增长情况所做的统计估算。

1. 保单 Bǎodān
在美国，指由制造商出具的、具有法律性约束的保证书，保证其产品符合某种质量标准，并在规定期限内负责更换有缺点的产品。又称担保。[参见"住房担保"]

2. 保证书 Bǎozhèngshū
在美国，由个人或组织出具的书面承诺，保证在另一个人或组织违约或不执行契约时，对其债务或义务承担责任。

H

Habitable room

Residential space used for living, sleeping, eating, cooking, or combinations of such activities but not including bathrooms, toilet compartments, closets, hallways, storage areas, laundry and utility rooms, and basement recreation rooms. A minimum floor area or ceiling height may be officially defined for such a room. The number of habitable rooms in the dwelling serves as an index for such purposes as determining rents.

Habitat

1. Area or region in which an animal or plant naturally lives or grows. See also ECOLOGY.
2. Dwelling place or habitation.

Halfway house

In the United States, a special group living arrangement for rehabilitation and support of persons recovering from drug or alcohol addiction or released from prison or jail. See also COMMUNITY FACILITY.

"Hard costs"

Expenses which are stable and relatively predictable. The precise meaning of the term varies according to context, for example, referring in commercial

居室 Jūshì

供起居、住宿、炊事或饮食用的居住空间,不包括浴室、厕所、壁橱、过道、储藏室、洗衣和杂用室、地下娱乐室等,对各类居室的最小建筑面积或净高一般有正式规定,居住单元内的居室数是确定房租的指标。

1. 生长地 Shēngzhǎngdì

动物或植物生长的地方。[参见"生态学"]

2. 住地 Zhùdì

居住的地方。

戒瘾教育所 Jièyǐn jiàoyùsuǒ

在美国,指为帮助吸毒或饮酒成瘾者戒掉恶习以及从监狱或拘留所中释放出的人而设立的一种特殊的集体住所。[参见"社区设施"]

硬费用 Yìngfèiyòng

一种稳定并可以预见的费用。该词的确切含义因不同用途而异,例如在商业预算中指租金或抵押贷款还款;在建筑工程中指人工费、材料费等可

budgeting to such costs as rent or mortgage payments or in construction engineering to definable items such as labor and materials.

Hazardous wastes

Waste matter or wastewater generated by an industrial or manufacturing process and potentially dangerous to the health of individuals who come into contact with it. Incidents of careless disposal of toxic wastes(chemical, nuclear, or other detritus)that may find their way into the groundwater or food chains have provoked concern about the safety of housing in the vicinity of disposal sites. See also NOXIOUS INDUSTRY.

Health care facility

A building, group of buildings, or area of a building designed to provide medical, nursing, or preventive health care services to the general public, employees, or a defined special group.

Heating, Ventilating, Air Conditioning (HVAC)

A system designed to provide a year-round indoor environment comfortable to human inhabitants by controlling such factors as temperature, moisture content of air, air quality, odor, and air circulation.

明确列出的项目。

有害废物 Yǒuhài fèiwù

由工业或制造过程产生的、对接触者的健康有潜在危害的废物或废水。处理上的疏忽可能导致有害废物（化学的、含核的或其他废品）进入地下或食物链。这个问题已引起在废物处理场地附近建造住房的顾虑。[参见"有害工业"]

保健设施 Bǎojiàn shèshī

为公众、职工或某一特定团体提供医疗、护理或保健服务而设计的一栋或一组建筑物，或建筑物中的一部分。

供暖、通风、空调 Gōngnuǎn, tōngfēng, kōngtiáo

通过对温度、湿度、空气质量、气味和空气流通等的控制，为居民提供全年舒适的室内环境而设计的系统。

Height zoning

In the United States, regulations for a particular zoning district designating the permissible height of buildings. See also BUILDING HEIGHT.

Heterogeneity

Diversity or lack of uniformity in composition, structure, or arrangement, as in the architectural style of a development or the makeup of a population. The opposite concept is homogeneity, a state characterized by uniformity.

Highest and best use

In the United states, an appraisal concept used to determine which of several uses of a particular piece of land will permit maximum development and prove most profitable in the foreseeable future. See also APPRAISAL.

High-rise building

See BUILDING HEIGHT.

High-rise/high density

Construction of multiple high-rise buildings on a site with little space or area between or around them, so that the residential density of the area is very high. The practice is common in cities with limited

建筑高度区划 Jiànzhù gāodù qūhuà

在美国，指在某一划定的区域内，对建筑物的允许高度的规定。[参见"建筑高度"]

多样性 Duōyàngxìng

在组成、结构或布置上的差异或不统一性，例如建筑的风格、人口的组成等都是千差万别的。其相反的概念是统一性或同一性。

最佳用途 Zuìjiā yòngtú

在美国，一种估价用的概念。用以确定一块土地的几种用途中，哪一种在可预见的将来会取得最大发展并获利最多。[参见"估价"]

高层建筑 Gāocéng jiànzhù

[见"建筑高度"]

高层高密度 Gāocéng gāomìdù

在一块土地上建造多栋高层建筑，而其间或四周只余较小空间或小块空地，该块土地上的居住密度因而很高。这种做法常在可建造用地比较紧张的城市中采用，例如香港。[参见"高层低密度"、

buildable land for construction, e.g., Hong Kong. See also HIGH-RISE/LOW DENSITY and LOW-RISE/HIGH DENSITY.

High-rise/low density

An approach to residential housing development that involves construction of one or several high-rise apartment buildings with many dwelling units, while leaving a large portion of the development site as open space or green areas. See also HIGH-RISE/HIGH DENSITY and LOW-RISE/HIGH DENSITY.

Historic American Buildings Survey

An archival program initiated in 1933 to document examples of American architecture of merit and historic value, using measured photographs and written records. The program is conducted through a cooperative agreement of the National Park Service, the Library of Congress, and the American Institute of Architects, and gives technical advice to U.S. federal agencies. See also CERTIFICATION OF HISTORIC BUILDINGS and HISTORIC PRESERVATION.

Historic District

In the United States, a portion of a city officially designated as historically significant, protecting that

"低层高密度"]

高层低密度 Gāocéng dīmìdù

住房建设的一种方法,即建造一栋或数栋供多户居住的高层公寓楼,而将建设用地的大部分留作空地或绿地。[参见"高层高密度"、"低层高密度"]

美国历史性建筑调查 Měiguó lìshǐxìng jiànzhù diàochá

美国于1933年开始执行的建立美国历史性建筑档案的计划,以精确的照片和文字记载将优秀的、具有历史价值的建筑收入档案。该计划是通过国家园林局、国会图书馆和美国建筑师学会达成的合作协议执行的,并向美国联邦政府机构提供技术性建议。[参见"历史性建筑保护证明"、"历史性建筑保护"]

历史性市区 Lìshǐxìng shìqū

在美国,指城市中被正式确定为具有历史意义的部分。该地区不得拆除,必须加以保护并提高受

area from demolition and enhancing the value of the protected properties. See also CERTIFICATION OF HISTORIC BUILDINGS, DISTRICT, HISTORIC PRESERVATION, NEIGHBORHOOD CONSERVATION, and RESTORATION.

Historic preservation

In the United States, restoration and reuse of old buildings, both as a means of preserving the country's architectural heritage and as an alternative to new construction. The National Historic Preservation Act of 1966 authorized the National Park Service to grant funds for preservation of historically significant districts, sites, and structures. See also CERTIFICATION OF HISTORIC BUILDINGS, HISTORIC AMERICAN BUILDINGS SURVEY, HISTORIC DISTRICT, NEIGHBORHOOD CONSERVATION, and RESTORATION. For the Chinese equivalent, see also CONSERVATION OF HISTORIC BUILDINGS.

"Holdback"

In the United States, a portion of a loan withheld by the lender or of payment held back by the owner until a particular requirement, e.g., completion of work, has been satisfied.

保护产业的价值。[参见"历史性建筑保护证明"、"区"、"历史性建筑保护"、"邻里保护"、"修复"]

历史性建筑保护 Lìshǐxìng jiànzhù bǎohù

在美国,指对旧建筑物的修复和重新使用,既是保护国家建筑遗产的手段,又是建造新建筑物的替代办法。1966年的美国《国家历史保护法案》授权国家园林局拨款保护具有历史意义的市区、遗址和建筑物。[参见"历史性建筑保护证明"、"美国历史性建筑调查"、"历史性市区"、"邻里保护"、"修复"。关于中国方面的相应词汇,参见"古建筑保护"]

暂扣款 Zànkòukuǎn

在美国,指在某项要求未满足以前,例如工程未完工,由贷款人暂扣的部分贷款或由业主暂扣的部分应付款项。

"Hold-harmless" provision

1. In the United States, contractual stipulation that one contracting party will not hold the other contracting party liable for damages.

2. In the United States, an arrangement to delay or prevent reduction of funds granted by the federal government to state and local governments. For example, under Title I of the Housing and Community Development Act of 1974, local governments may receive funds to close out grant programs that have been superseded and to phase in new programs.

Home improvement

In the United States, maintenance, upgrading, or remodeling of housing by individual home owners, either by their own labor or through contracting of work to small contractors. Individual home improvement has spawned a whole industry that produces and distributes partially finished and prefabricated building materials for the use of "do-it-yourselfers" through retail outlets. See also ADDITION, BUILDING ALTERATION, SELF-HELP HOUSING, and SWEAT EQUITY.

Home warranty

A guarantee that insures against structural defects, as

不受损条款 Bùshòusǔn tiáokuǎn

1. 在美国，指契约中的一项条款，规定订约的一方对另一方的受损不负赔偿责任。

2. 在美国，指延缓或免于执行联邦政府对州和地方政府削减拨款计划的规定。例如1974年的《住房和社区发展法》第一条中规定，地方政府可按已被取代的旧拨款计划继续取得拨款，直至旧计划执行完毕，开始新计划。

住房改善 Zhùfáng gǎishàn

在美国，指房主自出劳力或包给小型承包商进行住房的维修、改善或改建。私人改善住房的活动使生产半成品预制房屋材料的行业发展起来，并通过零售商店为自建者服务。[参见"加建工程"、"改建工程"、"自建住房"、"改进后净增值"]

住宅担保 Zhùzhái dānbǎo

对新购置的住房的结构及电线、水管、暖气、空

well as defects in electrical wiring, plumbing, heating, and air conditioning of a newly purchased home. The warranty is provided by a private company, e.g., the Home Owners Warranty Corporation, which is commissioned by a builder or seller to inspect and insure the home. See also GUARANTY.

Homesteading

1. In the United States, farming by individual families on a very small scale. Previously, under the Homestead Act of 1862, settlers could claim tracts of public land by living on and cultivating them. Under such an arrangement, the house and adjoining land are referred to as a homestead.

2. In the context of urban revitalization in the United States, homesteading involves transferring publicly owned, abandoned property in deteriorating neighborhoods at a token cost to individuals or families that agree to repair, occupy, and maintain the property. See also URBAN REVITALIZATION.

Homogeneity

See HETEROGENEITY.

Hospice

In Great Britain and the United States, a medical facility for care and support of the terminally ill.

气调节系统均无缺陷的担保。此种担保由私营公司提供,例如"房主担保公司"受营造商或卖主委托,检查住宅并提供担保。[参见"保单"]

1. 定居耕种 Dìngjū gēngzhòng

在美国,指由单独家庭经营的很小规模的耕种。早先,根据1862年的定居耕种法案,定居者可以要求得到他所居住和开垦的那一片公共土地,在这种情况下,其住房与四周土地称为家园。

2. 公有房地产转让 Gōngyǒu fángdìchǎn zhuǎnràng

在美国的城市复苏中,指在破落居住区中,把荒废的公用房地产,以象征性的少量代价出售给同意进行修理、使用和维护的个人或家庭。[参见"城市复苏"]

同一性 Tóngyīxìng

[见"多样性"]

晚期病人护理院 Wǎnqī bìngrén hùlǐyuàn

在英国和美国,指为护理和支持晚期病人而设立的医疗设施。

Household

1. In China, the registered inhabitants of a dwelling holding a common residence certificate issued by the police substation. See also FAMILY, RESIDENCE CERTIFICATE, RESIDENT REGISTRATION, and SINGLE-GENERATION HOUSEHOLD.

2. In the United States, a person or persons living together in one dwelling. A household may be classified for census purposes as a family or non-family household, depending on whether or not the individuals are related. The person in whose name the dwelling is owned or rented is designated the householder. See also FAMILY.

Household separation

In China, the withdrawal of persons from a household to form a new house-hold with another residence certificate.

Household with inconvenient space distribution

See HOUSING SHORTAGE.

Housing

Shelter or lodging for individuals and families, in the form of a house or dwelling unit in a multiunit building, or the provision of such shelter or lodging. Housing may be offered for purchase or rent, may be

户 Hù

1. 在中国,住在一处、领有一个由当地派出所颁发的户口簿的居民,称为一户。[参见"家庭"、"户口簿"、"户口登记"、"无子女户"]

2. 在美国,指一个住所内的一个或几个人。按人口普查目的,视他们是否有亲属关系而可划分为家庭住户或非家庭住户。住宅登记注册的所有者或租赁者为户主。[参见"家庭"]

分户 Fēnhù

在中国,指从原有的户中分出新户,另立户口簿。

居住不方便户 Jūzhù bùfāngbiàn hù

[见"住房短缺"]

住房 Zhùfáng

指供个人或家庭居住的单栋住宅或单元楼内的一个居住单元,亦可泛指提供住处。住房可购买或租赁,可由住户自行建造,或专为一个家庭建造,亦可由政府以低价提供。在美国,大约 2/3 的住

constructed by or for an individual family, or may be provided by the State at a nominal cost. In the United States, about two-thirds of all housing is owner-occupied.

Housing allowance

In the United States, a housing subsidy or voucher granted to lower income families which are renting apartments in the private market. The family pays a predefined portion of its income (usually one-quarter) as rent, and the government pays the difference between this amount and the fair market rent for the dwelling. Housing allowances may also be paid under other arrangements to employees of the military, the foreign service, or private companies who are required to live in high-cost areas. See also ASSISTED HOUSING, MARKET RENT, and SUBSIDY.

Housing association

In the United States, a private, nonprofit citizens' organization which seeks to ensure housing quality in the community through public relations, research, and advisory activities.

Housing authority

In the United States, a department or agency

房为住户自有。

住房津贴 Zhùfáng jīntiē

在美国,指政府给予低收入家庭在市场上租用公寓的住房补贴或住房补贴凭证。低收入家庭以其收入的一部分(常为1/4)用于租金,而与公平市场租价的差额则由政府补贴。对军队、外事机构或私营公司中必须住在高价地区的雇员,政府也可根据条件给予住房津贴。[参见"公助住房"、"市场租价"、"补贴"]

住房协会 Zhùfáng xiéhuì

在美国,指一种私人的、非营利性的市民组织,其目的是通过公众关系、研究和建议活动,以期保证社区中的住房质量。

住房主管部门 Zhùfáng zhǔguǎn bùmén

在美国,指政府设立的房管机关或部门,主持执

established by a government at some level to conduct programs ensuring that low-income families are housed in dwellings of at least minimum quality. There are approximately 2,500 housing authorities in the United States.

Housing census

See HOUSING SURVEY.

Housing code

See BUILDING CODES AND STANDARDS.

Housing commercialization

In China, a recent policy under which urban housing may be constructed and sold as a commodity to be used by both individuals and organizations instead of being unconditionally financed and allotted by the State.

Housing completions

The number of housing units that have been completed by the builder and officially certified as ready for occupancy. The number of housing completions is one of the measures of housing production. See also HOUSING PRODUCTION.

Housing conditions

The availability or the state of repair and maintenance

行住房计划,以保证低收入家庭至少能获得起码质量的住房。美国大约有 2500 个这种机构。

住房普查 Zhùfáng pǔchá

[见"住房调查"]

住房法规 Zhùfáng fǎguī

[见"建筑法规与标准"]

住房商品化 Zhùfáng shāngpǐnhuà

中国有关城市住房的政策,即城市住房可作为商品生产,向个人或单位销售,用以代替住房由国家投资兴建和无偿分配的做法。

住房竣工量 Zhùfáng jùngōngliàng

承包者建成并经正式验收可交付使用的住房套数,住房竣工量是统计住房产量的主要指标之一。[参见"住房建造量"]

住房状况 Zhùfáng zhuàngkuàng

住房的供应情况或修理与维护状况。[参见"住房

of residential dwellings. See also HOUSING SURVEY.

Housing cooperative

In the United States, a multiple ownership arrangement for real property in which a cooperative or business trust holds title to an apartment house and grants shareholders occupancy rights to dwelling units. Unlike condominiums, the separate units in a housing cooperative are not owned by individual occupants, and sales of occupancy rights are subject to approval of other shareholders. Also called a coop. See also COMMON-AREA MAINTENANCE, CONDOMINIUM, and MUTUAL HOUSING.

Housing demand

In the United States, the number of housing units that consumers are willing and able to purchase or rent at current price or rent levels. Also called effective demand. See also HOUSING MARKET ANALYSIS, HOUSING NEED, and HOUSING SUPPLY.

Housing estate

In China, a constituent part of a residential district, generally consisting of several housing groups that accommodate a total of 8,000 to 10,000 inhabitants, and are provided with public facilities

调查"]

住房合作社　Zhùfáng hézuòshè

在美国，指由多人共享房地产所有权的组织形式。公寓楼的产权归合作社或商业信用社所有，而各股东仅有使用居住单元的权利。与住户自有公寓不同，住房合作社中各单元的所有权不属于各住户。居住权的出售必须得到其他股东的同意。[参见"公用部分的维护"、"住房自有公寓"、"共有住房"]

住房需求　Zhùfáng xūqiú

在美国，指消费者愿意并有能力按当时的市价或租金水平购买或承租的居住单元总数。又称"有效需求"。[参见"住房市场分析"、"住房要求"、"住房供应量"]

居住小区　Jūzhù xiǎoqū

在中国居住区中，由若干住宅组团组成的、居住人口为8000—10000人的住区。区内除住房外，尚有公共服务设施及绿地等。[参见"住宅组团"、"居住区"]

and green spaces. See also HOUSING GROUP and RESIDENTIAL DISTRICT.

Housing group

In China, a group of residential buildings that constitutes a subdivision of a housing estate in a newly planned residential district. Each housing group, encompassing 700 to 800 households, is generally under the administration of one neighborhood committee. See also HOUSING ESTATE, NEIGHBORHOOD COMMITTEE, and RESIDENTIAL DISTRICT.

Housing management

Administration and control of dwelling units, encompassing tenant relations, rent collection, and building maintenance operations. In the United States, the term is usually applied to supervision of publicly assisted housing projects to ensure that government guidelines are properly applied. See also ASSISTED HOUSING and PUBLIC HOUSING.

Housing management office

See MUNICIPAL HOUSING MANAGEMENT BUREAU.

Housing market analysis

Systematic statistical examination of supply and

住宅组团　Zhùzhái zǔtuán

在中国，指新建居住小区内按规划布局建设成组的住房建筑群，一般以一个居委会管理700—800户的规模为一组。[参见"居住小区"、"居民委员会"、"居住区"]

住房管理　Zhùfáng guǎnlǐ

对住房的管理包括租户关系、收租和建筑物维修工作。在美国，本词常指对公助住房的监督，以确保政府的方针得以遵照贯彻。[参见"公助住房"、"公共住房"]

房管所　Fángguǎnsuǒ

[见"市房管局"]

住房市场分析　Zhùfáng shìchǎng fēnxī

对整个住房市场或一个大的地区内近期住房供求

demand relationships for the immediate future within an entire housing market sector or a major geographic area. See also HOUSING DEMAND and HOUSING SUPPLY.

Housing need

Requirements of the population for housing, either at the basic shelter level or according to a particular set of standards. This index is very general, and estimates vary greatly from country to country and even region to region. See also HOUSING DEMAND and HOUSING SHORTAGE.

Housing output

See HOUSING PRODUCTION.

Housing preference

Type of housing sought by the public. The long-standing preference of the American home buyer has been the single-family house on a large lot. However, increases in land, energy, and building costs, modifications in family constitution, and changes in attitudes and values have made high density alternatives such as condominiums and townhouses increasingly attractive. See also DETACHED HOUSE and LIFESTYLE.

关系所做的有系统的统计调查。[参见"住房需求"、"住房供应量"]

住房要求 Zhùfáng yāoqiú

指居民对于住房的要求,或仅按栖身之处的起码要求,或根据某些特定的标准而论。这一指标非常笼统,对其的估计因不同国家甚至不同地区而差异甚大。[参见"住房需求"、"住房短缺"]

住房产出量 Zhùfáng chǎnchūliàng

[见"住房建造量"]

住房选择 Zhùfáng xuǎnzé

指公众所喜爱的住房类型。长期以来,美国的购房者喜欢建在大基地上的单幢独户住宅。但由于土地、能源和建筑造价的提高,家庭构成的变化,观点和社会准则的改变,购房者的兴趣又转向高密度的住户自有公寓和市镇联立式住宅。[参见"独立式住宅"、"生活方式"]

Housing production

The construction rate of dwelling units for a particular area in a given time period. Housing output is commonly expressed as the number of housing starts, the number of housing units under construction, the number of housing completions, and the number of units occupied. The number and size of manufactured housing shipments, as well as the number of building permits issued are additional measures of housing production. See also CONSTRUCTION PUT-IN-PLACE, HOUSING COMPLETIONS, and HOUSING STARTS.

Housing program

1. In China, a program of housing development over a fixed number of years with yearly indices of residential districts scheduled for construction work, the amount of housing to be built, and the living standard to be achieved.
2. In the United States, this general term can mean any program for the construction, maintenance, or provision of housing, whether in the public or private sector.

Housing shortage

An inadequate supply of dwellings or of a particular type of dwellings suitable in size or quality to meet

住房建造量 Zhùfáng jiànzàoliàng

指在某地区的某一时期内,居住单元的建造速度。住房产出量通常以居住单元的开工量、施工量、竣工量及住用量来表示。预制住房的装运量和施工执照的颁发量也都是衡量住房生产的指标。[参见"建造量"、"住房竣工量"、"住房开工量"]

住房建设规划 Zhùfáng jiànshè guīhuà

1. 在中国,指制订出若干年内分年度建造住房的规划,包括居住区和住房建造的数量以及居住水平等多项指标。

2. 在美国,泛指公家或私人建设、维修或提供住房的计划。

住房短缺 Zhùfáng duǎnquē

指住房的供应不能满足全部人口对某类住房的大小或质量需要。在中国,有三种住房欠缺的情况:

the demand by all segment of the population. China records shortages for three types of households: 1) unallotted household – household for which no housing has been allotted because no units are available; for example, newly married couples may have to live in their parents' units that are already overcrowded; 2) over-crowded household – a household which lives in a dwelling with a floor area below the average minimum standard; and 3) household with inconvenient space distribution, e.g., separate bedrooms are not available for male and female children. In the United States, measurement of housing availability is usually based on the vacancy rate and on the number of units for sale or rent, by price of unit. Housing shortage is often measured by housing type and price range, as there may be a surplus in one price range and a shortage in another, depending in each category. See also HOUSING NEED, SHARED OCCUPANCY, SUBSTANDARD HOUSING, and VACANCY RATE.

Housing starts

The number of housing units for which building permits have been issued and construction has begun, commonly used in the United States as an economic indicator. See also HOUSING PRODUCTION.

①无房户,指因住房欠缺而未分配到住房的户,例如新婚夫妇因未分配到住房不得不与父母挤住在一处;②居住拥挤户,指居住面积低于某一地区的平均最低标准的户;③居住不方便户,指住房条件不便分居的户,例如不同性别的子女无法分室居住等。在美国,住房短缺与否,常以空房率、出售或出租的单元数量及价格来衡量,并常按住房类型和房价范围分类,有时某一价格范围内的住房短缺,而另一价格的住房却有余,因而各类住房短缺与否因市场供需情况而异。[参见"住房要求"、"合住"、"不合标准住房"、"空房率"]

住房开工量 Zhùfáng kāigōngliàng

获得开工执照并已开工的新住房的数量。在美国是一项经济指标。[参见"住房建造量"]

Housing stock

The quantity of existing housing units in a geographic area. Also referred to as housing inventory. See also HOUSING SRUVEY.

Housing supply

The number of dwelling units of various types available for sale or rent in a given area at a specific time. See also HOUSING DEMAND and HOUSING MARKET ANALYSIS.

Housing survey

In the United States, a study undertaken to obtain complete, accurate, and up-to-date information on the housing situation. The principal objective is to compare the size distribution of dwellings with the size distribution of households. At the national level, the Bureau of the Census periodically conducts the American Housing Survey, formally called the Annual Housing Survey. See also HOUSING CONDITIONS, HOUSING STOCK, and SIZE DISTRIBUTION OF HOUSEHOLDS.

Housing tenure

See TENURE.

Housing voucher

See HOUSING ALLOWANCE.

现有住房量 Xiànyǒu zhùfángliàng

在一个地区内当前所有的住房总量,亦称住房拥有量。[参见"住房调查"]

住房供应量 Zhùfáng gōngyìngliàng

在某地区的某一时期内,可供出租或出售的各类住房的套数。[参见"住房需求"、"住房市场分析"]

住房调查 Zhùfáng diàochá

在美国,指为取得完整、准确和最新的住房信息而进行调查研究。其主要目的是将户型比与户室比作出比较。在国家层面,美国人口调查局定期指导全美国的住房调查,该词正式名称为"年度住房调查"。[参见"住房状况"、"现有住房量"、"户型比"]

住房占用权 Zhùfáng zhànyòngquán

[见"占用权、占用期"]

住房补贴凭证 Zhùfáng bǔtiē píngzhèng

[见"住房津贴"]

Human ecology

The study of movements and settlement of human populations as influenced by their natural, social, and cultural environments. See also ECOLOGY and URBAN ECOLOGY.

Human resources

The pool of person with particular capabilities, whether general or expert, available to fill labor needs.

Human scale

Design of exterior and interior features of houses and arrangement of houses, streets, and open spaces in a manner both pleasing and comfortable to human inhabitants. Urban buildings of great size and bulk may be perceived as out of human scale unless such factors as amenities, space, and sunlight are considered in their design.

Human services

In the United States, activities undertaken by governmental agencies and private organizations to promote public welfare, as well as mental and physical health, especially for children, the elderly, and the disadvantaged. Also known, although less commonly, as social services.

人类生态学 Rénlèi shēngtàixué

研究人类在自然、社会和文化环境影响下的迁移和定居的科学。[参见"生态学"、"城市生态学"]

人力资源 Rénlì zīyuán

具有某种一般或专业能力,能满足劳动需要的备用人员。

人类尺度 Rénlèi chǐdù

指住宅的室内外特征以及住宅、街道与室外空间的布局设计应使人类感到愉快和舒适。大尺寸和大体积的城市建筑物,除非在设计中已经考虑到它的舒适性、空间和阳光等因素,否则就会被认为超出了人类尺度。

人类服务 Rénlèi fúwù

在美国,指由政府部门或私人组织为增进公共福利、促进身心健康(尤指老人、小孩、残疾者)而从事的活动和工作,又称社会服务。

"Hutong"

In China, a narrow street between buildings or groups of buildings, which may be a dead end. The "hutong" is similar to the "alley" in the United States. Also called "li" in northern China, and a "nongtang" or "nong" in southern China.

HVAC

See HEATING, VENTILATING, AIR CONDITIONING.

胡同 Hútòng

在中国，通常指建筑物或建筑群之间的小路。有时路的尽头可能被挡住，称死胡同，与美国的小巷相似。胡同在北方又称"里"，在南方称"弄堂"或"弄"。

供暖、通风、空调（缩写） Gōngnuǎn, tōngfēng, kōngtiáo (suōxiě)

［见"供暖、通风、空调"］

I

Impact

The sum total of positive or negative effects of any policy, procedure, or project on systems, environment, quality of life, or other conditions. See also ENVIRONMENTAL IMPACT STATEMENT, EXTERNALITIES, and SIDE EFFECT.

Incentive zoning

In the United States, a zoning system that awards bonuses to developers for providing desirable public benefits, for example, higher permitted densities or special street arrangements may be granted in exchange for inclusion of plazas or access to transit stops in developments. See also BONUS INCENTIVES FOR DEVELOPMENT.

Income stream

See CASH FLOW.

Indirect expenses

Expenses which are incurred for but not chargeable to a particular construction project and must be allocated according to a certain formula to various items of the project, i.e., all administrative expenses incurred in the course of organizing and managing the construction work. Under the Chinese system, indirect expenses are charged according to a certain

影响 Yǐngxiǎng

任何政策、程序或计划对于各种系统、环境、生活条件或其他条件所造成的积极或消极效果的总和。[参见"环境影响报告"、"外在副作用"、"副作用"]

鼓励性区划 Gǔlìxìng qūhuà

美国的一种区划体系,其中规定对于愿意提供公益设施的开发者,例如在社区开发中设置广场或通向换车站的通道,可以获得特准的较高建筑密度或特殊街道布局等优待。[参见"开发奖励"]

收入流量 Shōurù liúliàng

[见"现金流量"]

间接费 Jiànjiēfèi

指不能或不宜直接计入而必须按一定标准分配到工程项目上去的费用,即为建筑安装或施工中在组织和管理上所需的各项费用。在中国,间接费按照直接费的一定百分比计取,又称施工管理费。[参见"建筑成本"、"直接费"、"独立费"、"管理费"]

percentage of the direct expenses. Also referred to as construction management fees. See also CONSTRUCTION COST, DIRECT EXPENSES, EXTRA EXPENSES, and OVERHEAD.

Individual business

In China, a small commercial or service business run by individuals with their own production systems and equipment, encompassing anything from street vendors to restaurants and various kinds of small service businesses. See also COTTAGE INDUSTRY.

Industrial-agricultural enterprise

In China, an association for mutual benefit between an industrial and an agricultural enterprise, generally an urban enterprise and a rural commune enterprise in closely related areas of production. See also AGRO-INDUSTRIAL-COMMERCIAL ENTERPRISE.

Industrial building

A factory building used for production of manufactured goods. According to the Chinese classification system, industrial buildings include all buildings used for production purposes. See also CIVIL BUILDING.

Industrial district

See DISTRICT.

个体户 Gètǐhù

在中国,指生产资料归个体所有,由个体经营的商业、服务业,包括小商贩、个体经营的饭馆等。[参见"家庭工业"]

工农联合企业 Gōngnóng liánhé qǐyè

在中国,指由在生产上有密切联系和共同利益的城市与农村社队的工业企业和农业企业联合组成的企业。[参见"农工商联合企业"]

工业建筑 Gōngyè jiànzhù

指用于工业生产的各种建筑物。在中国,工业建筑包括用于生产的各种建筑物。[参见"民用建筑"]

工业区 Gōngyèqū

[见"区"]

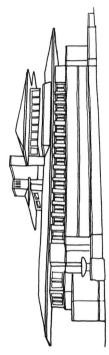
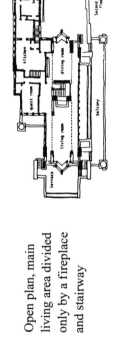

Prairie Style
Robie House, 1908-1909
Chicago, Illinois
(Architect: Frank Lloyd Wright)
名建筑师赖特设计的住宅

Open plan, main living area divided only by a fireplace and stairway

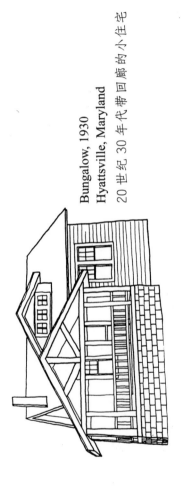

Bungalow, 1930
Hyattsville, Maryland
20 世纪 30 年代带回廊的小住宅

Typical features:

Wide eaves, exposed rafters

Long, low-lying roof

Heavy porch supports

Industrial modernization investment

In China, investment in renovation, primarily of buildings for production purposes. The required financial resources may be drawn from depreciation funds established by the enterprises. Funds may also be derived from State budgetary appropriations for renovation; the enterprises' own funds; or domestic or foreign loans for technical remodeling. See also FIXED-ASSET INVESTMENT, INDUSTRIAL MODERNIZATION PROJECT, and INVESTMENT IN CAPITAL CONSTRUCTION.

Industrial modernization project

In China, a project approved by the State or the responsible authorities, with design documents, to renovate or to retrofit existing fixed assets or for technical remodeling of existing facilities by using investment funds for industrial modernization. See also INDUSTRIAL MODERNIZATION INVESTMENT.

Industrial park

In the United States, a parcel of land, frequently located on the outskirts of a city, and typically under single management but designed for use of a group of businesses and industries. Industrial parks usually house light industry and often provide special services and facilities to tenants. Emphasis is placed

更新改造措施投资 Gēngxīn gǎizào cuòshī tóuzī

在中国,指利用企业基本折旧基金、国家更新改造预算拨款、企业自筹资金以及国内外技术改造贷款等,对原有项目,主要是生产性建筑,进行更新改造的投资。[参见"固定资产投资"、"更新改造措施项目"、"基本建设投资"]

更新改造措施项目 Gēngxīn gǎizào cuòshī xiàngmù

在中国,指对原有的固定资产进行更新改造的工程措施项目。这些项目,一般都经国家或主管部门批准,给予更新改造的投资,并具有独立的设计文件,其内容也包括对房屋原有设施的技术改造。[参见"更新改造措施投资"]

工业园 Gōngyèyuán

在美国,通常指位于都市边缘的一块独家管理的土地。其中安排若干商业与工业,尤其是轻工业,并常对租用者提供特殊服务和设施。区内注意环境舒适、风景宜人,常受到美观方面的规范控制。

on amenities and attractive landscaping, frequently under regulation of aesthetic controls.

Industrial revenue bond

In the United States, a type of bond issued by a municipality to finance construction of an industrial or commercial facility for rental by a private company. The income from interest on the bonds is exempt from federal taxation. See also BOND.

Industrialized building system

A building process using standardized, mass-produced components and allowing total integration of industrialized production, transportation, and assembly techniques. Industrialized housing, i.e., housing units or portions of housing units such as walls, rooms, and plumbing fixtures, may be produced in this manner. See also MANUFACTURED HOUSING, MODULAR COORDINATION, MODULAR HOUSING, and MODULE.

Industrialized housing

See INDUSTRIALIZED BUILDING SYSTEM.

Infill development

The construction of new structures in the unused area between or within existing built-up areas to make maximum use of infrastructure and transportation.

工业债券 Gōngyè zhàiquàn

在美国,指由市政当局发行的一种债券,用以筹措资金,供建造出租给私人企业的工业或商业设施。工业债券的利息收入可以免除联邦所得税。[参见"债券"]

工业化建筑体系 Gōngyèhuà jiànzhù tǐxì

采用大批量生产标准化的建筑构件,并把工业化的生产、运输和安装等全部技术结合起来的一种建筑施工方法。工业化住房不论整幢还是其中的一部分,例如墙、房间、卫生设备等,均可用这种方法生产。[参见"工厂预制住房"、"模数协调"、"模数制住房"、"模数"]

工业化住房 Gōngyèhuà zhùfáng

[见"工业化建筑体系"]

填空性建设 Tiánkòngxìng jiànshè

在建成区之间或之内尚未使用的地段上所进行的新建工程,以充分利用基础设施和运输条件。[参见"零星建设"、"见缝插针"]

See also PIECEMEAL DEVELOPMENT and SQUEEZE-IN DEVELOPMENT.

Informal sector

Term used in economics to designate the portion of an economy's labor force, especially in developing countries, involved in small-scale, labor-intensive work such as tailoring, cooking, and shoe repair. While such workers are not reflected in employment statistics because they are considered unemployed or underemployed, they are generally highly productive, contributing significantly to national income.

Information science

The field of study and body of knowledge concerned with the classification, storage, and retrieval of written documents or information recorded in some other form. Information science also involves design of manual, computerized, or electronic systems to facilitate accessing and exchanging such information.

Infrastructure

Systems of physical structures on or under the ground to provide services, access, or convenience in urban areas, that is, roads, water and sewer lines, curbs and gutters, telephone and electric lines, and so forth. The term infrastructure is sometimes also broadly applied

非正规行业 Fēizhèngguī hángyè

经济学用语。指经济部门中的一部分劳动力,尤指发展中国家从事小规模劳动密集型的行业,如缝纫、烹饪、修鞋等。由于这些工人被认为是失业或未充分就业,故在就业统计中反映不出来,但他们的生产力很强,对整个国民收入有明显贡献。

信息科学 Xìnxī kēxué

专用研究文献或其他信息载体的分类、存储和检索的科学。信息科学也包括用手工、计算机或电子化等操作体系,进行信息的存取和交换。

基础设施 Jīchǔ shèshī

指城市地区中在地上或地下提供服务、通道或便利的实体结构,如道路、给排水管道、路边石和边沟、电话与电力线路等。本词有时也泛指物质及社会性的基础结构,包括城市社区生活中的一些组成部分,如运输与通信系统、购物设施、住

to both physical and social infrastructure, including social services and facilities which are an integral part of life in an urban community, e.g., transportation and communication systems, shopping facilities, housing, schools, and recreation facilities. See also COMMUNITY FACILITY, PUBLIC UTILITIES, PUBLIC WORKS, and SERVICED LAND.

Inner city

In the United States, a central section of a city usually characterized by deteriorating, substandard housing, low-income population, and high crime rates, but often also possessing universities, historic buildings, hospitals, shopping areas, parks and other amenities used by residents of the whole metropolitan area. See also GHETTO.

Inner suburban district

See SUBURB.

Inspector

See CONSTRUCTION INSPECTOR.

Institutional framework

The interrelated organizations and agencies that provide a basis for development and implementation of policies, programs, and projects in a particular field.

房、学校和娱乐设施等。[参见"社区设施"、"市政公共设施"、"公用设施"、"有设施用地"]

城内衰落区 Chéngnèi shuāiluòqū

在美国,指市区中心住房破旧、不合标准、居民收入低、犯罪率高的地区。但区内也有大学、历史性建筑、医院、购物区、公园以及供全城居民使用的其他设施。[参见"少数人种聚居区"]

近郊区 Jìnjiāoqū

[见"郊区"]

检查员 Jiǎncháyuán

[见"施工检查员"]

相关机构网 Xiāngguān jīgòuwǎng

在某一领域中,为某些政策、计划和项目的发展与实施提供基础而相互关联的组织和部门。

Interior design

Originally, professional planning and coordination of the style, colors, and layout for the interior of a building and its furnishings. In parts of the United States, interior designers have become involved in every aspect of space planning and now often constitute a separate but integral part of large architectural firms. See also SPACE PLANNING.

Intermediate technology

A name coined by E.F. Schumacher for the techniques and technological processes midway between capital-intensive western technologies and simple, labor-intensive methods used in developing countries. See also ALTERNATIVE TECHNOLOGY.

Interstate Highway System

In the United States, a 41,000-mile transcontinental limited-access highway network that was constructed primarily with federal government funds (90 percent), the remainder deriving from the states. Maintenance is primarily the responsibility of the states, although some federal funds, including special tax revenues, are used for this purpose. See also EXPRESSWAY.

Investment in capital construction

In China, funds used for carrying out capital

室内设计 Shìnèi shèjì

原指为建筑物内部及其陈设所做的关于风格、颜色及布置上的专业设计与协调。在美国某些地区,室内设计已涉及空间规划的各个方面而常成为大型建筑设计事务所业务的组成部分。[参见"空间布局"]

中间技术 Zhōngjiān jìshù

由 E.F. 舒马彻提出的专用术语,指介于资本密集型的西方技术与发展中国家所用的简单的劳动密集型技术之间的技术和工艺流程。[参见"替换技术"]

州际公路网 Zhōujì gōnglùwǎng

在美国,指主要由联邦政府投资兴建的横跨大陆的、限制入口的公路网,总长 41000 英里。联邦政府出资 90%,其余经费由各州提供。公路的保养则由各州负责,但联邦政府亦有部分基金,包括专项税收收入,用于此项目。[参见"高速公路"]

基本建设投资 Jīběn jiànshè tóuzī

在中国,指用于进行基本建设的资金,是工程项

construction, encompassing the total construction payment for the project, including all the expenses for construction work, the procurement of equipment, and other expenditures, e.g., design and surveying fees, land acquisition fees, demolition and restoration expenses, and training expenses. Financial resources for capital construction include State budgetary appropriations; outside funds raised by individual ministries, local governments, or enterprises; domestic and foreign loans; and other specific funds. The amount of investment in capital construction is the work quantity expressed in terms of a monetary unit and represents the index that defines the scope of the project. See also CAPITAL CONSTRUCTION, DOMESTIC LOAN, FIXED-ASSET INVESTMENT, FOREIGN LOAN, INDUSTRIAL MODERNIZATION INVESTMENT, ORGANIZATION-RAISED FUNDS, SCOPE OF CONSTRUCTION PROJECT, and STATE INVESTMENT.

Investment tax credit

In the United States, deduction from taxable income allowed on tangible personal property or other property used for business, production, or provision of certain services as an inducement to economic growth and job creation.

目的全部建设费用。基本建设投资包括建筑安装工程的费用、设备、工具、器具的购置费用，以及其他基本建设费用，如设计勘察费、征用土地费、拆迁补偿费、培训费等。基本建设投资的资金来源，包括国家预算内的基本建设拨款、自筹资金、国内外基本建设贷款以及其他专项资金。基本建设投资额是以货币形式表现的基本建设工作量，是衡量基本建设规模的综合性指标。[参见"基本建设"、"国内贷款"、"固定资产投资"、"国外贷款"、"更新改造措施投资"、"自筹资金"、"建设项目规模"、"国家投资"]

投资减税额 Tóuzī jiǎnshuì'é

在美国，为鼓励将有形动产或其他资产用于企业、生产或某些服务业，以利于经济增长和增加就业机会，对其应征所得税的减让额。

J

"Jerry-built"

Constructed in a cheap, flimsy, makeshift manner with poor quality materials.

Job assignment notice

In China, the form for organizing construction teams to perform the construction work. The job assignment notice states the nature of the work, responsibilities, work quantity, the construction norm, deadlines, and quality and safety requirements. Blank spaces are provided for recording the actual work completed. The form is used as the basis for accounting and statistics.

Job captain

In the United States, a member of the designer's staff normally responsible on a given project for the preparation of drawings and their coordination with other documents.

Job displacement

Shift of positions, and consequently of a labor force, to a different region because of relocation of a business or industry. See also DISPLACEMENT OF FAMILIES and RELOCATION.

Job superintendant

In the United States, a contractor's representative at a

简陋搭建 Jiǎnlòu dājiàn

用廉价材料盖劣质房屋。

工程任务单 Gōngchéng rènwùdān

在中国,指安排施工队执行施工任务的单据,是核算和统计的依据。工程任务单上包括工作内容、任务、数量、施工定额、完成期限以及对质量和安全的要求。单内还列有供填写实际完成任务的空栏。

设计组长 Shèjì zǔzhǎng

在美国,对某一工程负责提供图纸以及图纸与其他文件的协调工作的设计人员。

就业转移 Jiùyè zhuǎnyí

由于工商业的迁址而引起就业岗位的转移,从而造成劳动力的转移。[参见"家庭迁移"、"搬迁"]

工段长 Gōngduànzhǎng

在美国,承包商驻工地的代理人,其职责是负责监

construction site who is responsible for supervision, coordination, and completion of work on a construction project, as well as for job safety.

Joint venture

A specific undertaking or project conducted by two or more persons, businesses, or organizations with equal control and liability, frequently to develop and market a new product.

督、协调与完成工程施工,也负责防止工程事故。

合资经营 Hézī jīngyíng

由两个或两个以上具有相等的管理权和责任的个人、商号或组织共同经营的企业或项目,通常从事新产品的开发和销售。

K

Kitchenette

In the United States, a room for food preparation, usually less than sixty square feet in area, containing a modular kitchen unit and designed for efficient space utilization in small apartments.

小厨房 Xiǎochúfáng

在美国,指在小套公寓内,为了有效地利用空间而设计的烹饪的地方,通常小于60平方英尺,包含一套符合模数的厨房设施。

L

Labor equilibrium

In China, a concept used by urban planners to plan a city's population growth, considering the size of the population, policy objectives for national economic and social development, and occupational areas in which resources are needed. The major objective of planning for labor equilibrium is to ensure an adequate supply of service workers to support primary workers, and appropriate goals for total population levels in relation to available space. See also NATIONAL ECONOMIC AND SOCIAL DEVELOPMENT PLANNING.

Labor population

In China, the total number of urban residents who are employed. See also PRIMARY POPULATION and SERVICE POPULATION.

Labor requirements per square meter

In China, the average amount of labor expended to construct a square meter of floor area on a construction project. See also NORM FOR ESTIMATING LABOR REQUIREMENTS.

Laboratory accreditation

See ACCREDITATION.

劳动平衡 Láodòng pínghéng

中国的城市规划工作者所采用的一种方法,即以人口的数量、国民经济与社会发展计划及各行各业所需人力资源为基础,测算人口的增长。其目的在于保证基本人口与服务人口的平衡以及可用土地范围内总人口的适当规模。[参见"国民经济与社会发展计划"]

劳动人口 Láodòng rénkǒu

在中国,指城市总人口中参加工作的人口。[参见"基本人口"、"服务人口"]

单方用工 Dānfāng yònggōng

在中国,指建设工程中每平方米建筑面积的平均用工量。[参见"劳动定额"]

实验室资格证明 Shíyànshì zīgé zhèngmíng

[见"确认"]

Land acquisition

1. In the United States, obtaining the title to land through the right of eminent domain, purchase, inheritance, gift, or other means. In the transaction, ownership of the land is transferred from one person or organization to another. See also ASSIGNMENT OF TITLE, COMPENSATION FOR LAND, and EMINENT DOMAIN.

2. In China, procurement of the right to use land currently used by state-owned or collectively owned organizations, or allotted as family land plots in urban, suburban, or rural areas, for the purpose of planned development. In such cases, the holder or holders of the right to use the land must be compensated. See also ACQUISITION APPROVAL, COMPENSATORY REPLACEMENT OF DEMOLISHED HOUSING, CROP LOSS COMPENSATION, and TRANSFER OF DOMICILE REGISTRATION.

Land banking

A practice in which land is purchased by the government in anticipation of future development requirements and is held in reserve until needed. Land banking is especially common In European countries as an alternative to eminent domain.

征地 Zhēngdì

1. 在美国，指通过国家征用、购置、继承、受赠或其他途径获得的土地所有权。在移交时，土地所有权由一方（个人或机构）转让给另一方。[参见"过户"、"土地补偿费"、"国家征地权"]

2. 在中国，指因建设需要而征用城市市区、郊区或农村中属于全民所有制或集体所有制单位的土地，或农民的自留地。征用一方需给予原土地使用者补偿费。[参见"拨地"、"原拆原建"、"农田青苗补偿"、"转户"]

地产储备 Dìchǎn chǔbèi

指政府预先购置地产，留存以供日后建设发展需要时使用的做法。此种做法在欧洲国家尤为普遍，用以代替国家征地权。

Land conservation

See CONSERVATION.

Land costs

In the United States, the amount of money required to purchase land for housing or other uses.

Land parcel

In the United States, a continuous plot of land of any size in the possession of one owner. For the purpose of subdivision development, land is parceled out into individual building plots or large parcels of land, as shown on a plot layout.

Land readjustment

A relatively new term that usually refers to the conversion of rural or semirural land to urban building sites or to the redevelopment of existing urbanized areas; also known as urban consolidation or urban land pooling. Essentially, land readjustment refers to a method by which a city government or another public body becomes the principal land developer. Titles to all land parcels within a designated area are "pooled," a site plan prepared, and the costs of constructing physical infrastructure and community facilities to service the site are calculated. The public developer builds the necessary

土地保护　Tǔdì bǎohù

［见"保护"］

地价　Dìjià

在美国，指为建住房或其他用途而购买土地的金额。

地块　Dìkuài

在美国，指由一个地产主所有、大小不同但统一使用的相互连接的一片地。为了满足土地划分后建设的需要，可将其划成若干块建筑用地，或划成几大块，以用地布置图表示。

土地调整　Tǔdì tiáozhěng

一种新的土地规划用语。指将农村或半农村土地改变成城市的建设场地或用于现有城市化地区的再开发。土地调整又称城市整顿或城市土地聚集。此方法的特点是使城市政府或其他公共团体成为主要土地开发者。其做法是将指定地区内全部地块的所有权集中，拟出建设场地的总平面图，并算出拟建基础设施和社区设施的费用，由开发者建造必备的设施以改善土地状况，然后出售部分设施用地以收回成本。剩下的土地，按原土地所有者占地的比例发还。如计算正确，则城市既可开发出新的市区，地产主还可以分得在原有产业附近的土地。虽然分回的土地面积较小，但经重

improvements and sells enough of the now serviced land to recover the costs. The remaining lots are then distributed to the original landowners on a proportionate basis. If the calculations are correct, the city can create new urbanized districts, while the original landowners receive lots as near their original holdings as possible. Although these lots are smaller than previously, their value has increased because they are now subdivided and serviced. An advantage is that the whole process of urbanizing land is carried out through manipulation of land titles with little or no expenditure of public funds. Countries with extensive experience in this land development method are Japan, West Germany, South Korea, and Australia.

Land reclamation

1. Restoration of land that is overworked, contaminated, flooded, or otherwise unserviceable to a condition in which it can be used for agriculture or other purposes.

2. Replanning of prematurely subdivided, lightly settled land that would otherwise never be properly used because of tax delinquencies, clouded titles, or poor development design.

3. Public acquisition of land unsuited for private development.

新划分,且已具备设施,故地价提高。这种做法的优点是通过控制土地所有权来实现土地城市化,而公家花费甚少,或不花。推行此法有经验的国家有日本、西德、韩国、澳大利亚等。

土地整治 Tǔdì zhěngzhì

1.将因过分使用、污染、洪水淹没而不能再使用的土地恢复到可用于农业或其他目的的状况。

2.将划分不当、草率占用的土地重新规划,不然这些土地将因欠税过多、所有权不清或开发设计不善而不能被正常使用。

3.由公家征用不适合私人开发的土地。

Land survey

In the United States, determination of the official boundaries of a plot of land or of a roadway using geometric and trigonometric measurement techniques. Also called a boundary survey.

Land tenure

See TENURE.

Land title

See TITLE.

Land use assessment

For land planning in China, the evaluation of a piece of urban land with regard to the feasibility of development, proposed use, and investment required, as well as the economic, social, and environmental effects on the ecological balance of the city.

Land use balance

In China, a proportional distribution of urban land among use categories, e.g., industrial, agricultural, and residential areas; roads; green spaces; and so forth. Achieving this goal requires determination of appropriate area distribution for respective usages and of the relative proportions of use areas, as well as preparation of a table of land use balance, from which

土地测量 Tǔdì cèliáng

在美国,指用三角或几何测量方法确定一块土地或道路的正式界线,又称"边界测量"。

土地占用权 Tǔdì zhànyòng quán

[见"占用权、占用期"]

地契 Dìqì

[见"产权证"]

城市用地评价 Chéngshì yòngdì píngjià

在中国的城市规划中,对城市土地开发的可行性、用途、需要投入的资金,以及对城市的社会、经济、自然生态平衡的影响所做的评价。

用地平衡 Yòngdì pínghéng

在中国的城市规划中,按土地的使用性质,将全市或局部地区划分为工业、农业、居住、道路、绿化等用地;确定各类用地的合宜面积分配与比例关系,编制用地平衡表,从而得出以人为单位的用地面积数,以资分析和比较。

land areas per capita are obtained for analysis and comparison.

Land use controls

Measures or restrictions intended to define the use of land for any purpose or type of development. Land use may be controlled in a variety of ways, e.g., by zoning or general environmental restriction, as well as by enforcing relevant regulations. See also AESTHETIC CONTROLS, TRANSFERABLE DEVELOPMENT RIGHTS, ENVIRONMENTAL STANDARDS, and ZONING.

Land use plan

In the United States, a plan for future development of a community or similar jurisdiction showing the size and location of areas to be used for residential, commercial, industrial, public, and other purposes. The land use plan often serves as the basis for specific zoning decisions.

Land use survey

In the United States, a critical examination of the various types of land development and use in a particular area, usually undertaken as part of a comprehensive planning effort. See also COMPREHENSIVE PLANNING and QUANTITY SURVEY.

土地使用控制　Tǔdì shǐyòng kòngzhì

指对用于各种建设的土地的一些限制性规定或措施。控制的办法有土地区划法、全面环境限制和执行有关规定等。[参见"美观控制"、"可转让开发权"、"环境标准"、"区划"]

土地利用规划图　Tǔdì lìyòng guīhuàtú

在美国,指对一个社区或类似范围内日后的发展所做的规划图,其中标明用于居住区、商业区、工业区、公共活动区或其他用途的土地的大小和位置。土地利用规划图常是确定具体区划的基础。

土地利用调查　Tǔdì lìyòng diàochá

在美国,指对某一地区内不同类型的土地开发及使用情况的详细检查,是总体规划工作的一环。[参见"总体规划"、"空地或空房调查表"]

Landscape architecture

The art and profession of planning, developing, or modifying land or natural scenery to achieve a desired purpose or effect and to improve the quality of human life. See also "BORROWED VIEW," DESIGN WITH NATURE, ENVIRONMENTAL DESIGN, "SCENIC FOCAL POINT," and "YIJING."

Large city

See CITY SIZE.

Layout plan

See SITE PLANNING.

"Lead time"

In American speech, the time before a planned event or project start which is necessary or desirable for making the required plans and preparations.

Leasable area

In the United States, the amount of space in a building that can be rented out. The gross leasable area refers to the total amount of floor space in a building that can be rented to tenants and is used in describing the property. Net leasable area designates

风景建筑学 Fēngjǐng jiànzhùxué

对土地或自然景观进行规划、开发和改善的艺术或专业,以达到预期的目的或效果,并改善人类的生活环境。[参见"借景"、"适应自然的设计"、"环境设计"、"对景"、"意境"]

大城市 Dàchéngshì

[见"城市规模"]

建筑布置图 Jiànzhù bùzhìtú

[见"建设场地规划"]

准备时间 Zhǔnbèi shíjiān

在美国用语中,指在一项有计划的活动或工程开始前,用来做出必要的计划和进行准备所需的时间。

可出租面积 Kěchūzū miànjī

在美国,指建筑物内可供出租的总面积。可出租毛面积指一栋建筑物内可出租的总建筑面积,用以说明一项产业的规模。可出租净面积不包括楼梯间、门厅、电梯井等,是用户实际可使用的面积。[参见"可出租面积比"]

the amount of building space actually usable by tenants, excluding such areas as stairways, halls, elevator shafts, and so forth. See also BUILDING EFFICIENCY.

Lease

In the United States, a contract between the owner of a building or land and another party stipulating the terms under which the tenant may occupy and use the property, as well as the rights and responsibilities of both lessee and lessor. A ground lease is a legal document for the rental of land over a specified number of years. See also LEASEHOLD, LEASEPURCHASE AGREEMENT, LESSEE,LESSOR, and RENTAL HOUSING.

Leasehold

In the United States, a rental arrangement for real or personal property for a fixed number of years. Under the arrangement, the lessee has the right to use the land or property, but the lessor continues to hold the title. See also LEASE.

Lease-purchase agreement

In the United States, an arrangement for renting buildings or other assets that allows the lessee to lease and eventually to exercise the option to buy

租约 Zūyuē

在美国,指房地产的业主(出租人)和另一方(承租人)所签订的契约,规定承租人可使用该产业,并规定双方的权利和义务。土地租约则是规定土地租赁年限的法律性文件。[参见"租赁"、"租赁购置协议"、"承租人"、"出租人"、"租用住房"]

租赁 Zūlìn

在美国,指在一定年限内租借动产或不动产的安排,其中规定承租人有权使用该土地或产业,但出租人继续保持其所有权。[参见"租约"]

租赁购置协议 Zūlìn gòuzhì xiéyì

在美国,指租赁房屋或其他固定资产的一种协议,其中规定承租人的租赁权,并允许最后有权选择是否以按月付款的方式购置该产业,已付租金可

the asset for a monthly fee. Rental payments may be applied to reducing the eventual purchase price. See also LEASE.

Lessee

In the United States, a person or organization granted use of land or property by another person or organization under a rental contract or lease; the tenant. See also LEASE and LESSOR.

Lessor

In the United States, a person or organization granting use of land or property to another person or organization under a rental contract or lease; the landlord. See also LEASE and LESSEE.

Level crossing

An intersection of a road and railway on the same level. Also called a grade crossing.

Leveraging

1. In the United States, use of public or private funds for investment in order to stimulate increased investment of funds from other sources. See also MULTIPLIER EFFECT and URBAN DEVELOPMENT ACTION GRANT.

2. In the United States, the use of borrowed funds

以抵充购置价款的一部分。[参见"租约"]

承租人 Chéngzūrén

在美国,指通过签订租约而获权使用他人或另一团体的土地或产业的个人或团体,也称租户。[参见"租约"、"出租人"]

出租人 Chūzūrén

在美国,指通过签订租约而将其土地或产业的使用权出让给另一方的个人或团体,也称地主或房东。[参见"租约"、"承租人"]

平交 Píngjiāo

道路或铁路在同一水平上的相交。

杠杆作用 Gànggǎn zuòyòng

1. 在美国,指用公共或私人资金投资来刺激和增加其他方面资金的投资。[参见"收入增殖作用"、"城市开发活动赠款"]

2. 在美国,也指利用借入资金来获取利润,在这种

to realize profit on an investment. In such an arrangement, the cost of borrowed money should be lower than the expected rate of return on the investment.

Liabilities

In the United States, monetary obligations or debts payable in money, property, or services, as opposed to holdings or assets. See also ASSETS.

Lien

In the United States, a legal claim on a property, usually resulting from its use as security for a loan. Upon sale of the property, the lien must be paid off before the owner receives money from the sale. See also MORTGAGE and SECURITY.

Life cycle costing

A method for determining the price of an item in relation to its total anticipated operating life. An item with an initially high purchase price may prove cost-effective over a long time period, because a lower cost, lower quality item may be replaced several times in the same period.

Lifestyle

The distinct living patterns, habits, and preferences of

情况下，借入资金的利息应低于投资的预期收益。

负债 Fùzhài

在美国，指货币性的义务或应以现金、产业或服务偿还的债务，是"资产"的反义词。[参见"资产"]

留置权 Liúzhìquán

在美国，指对于一项产业的法定索赔权，常因以该产业作为借款的抵押品而产生。当出售该产业时，必须将留置抵押的债务偿清，业主方能取得售款。[参见"抵押贷款"、"担保品"]

全寿命费用计算 Quánshòumìng fèiyòngjìsuàn

一种结合产品的预期使用年限计算产品价格的方法。该方法认为，购置价格高的产品从长远看，其经济效益可能更好，因为价格低而质量不好的产品，在相同时间内可能要更换多次。

生活方式 Shēnghuó fāngshì

指个人或集体所特有的不同的生活形态、习惯和

any particular individual or group of individuals. See also HOUSING PREFERENCE.

Living floor area

In China, the total net floor area of living rooms and bedrooms in a dwelling unit. In urban planning, gross floor area is more commonly used. See also DENSITY OF LIVING FLOOR AREA, RESIDENTIAL FLOOR AREA, and USABLE FLOOR AREA.

Loan terms

Conditions for repayment of a sum of money borrowed, as defined by a loan agreement.

Loan-to-value ratio

In the United States, the relation of the loan amount for a property to the appraised value of that property. The loan-to-value ratio is one factor that affects the rate of interest on the loan; the higher the loan-to-value ratio, the higher the interest rate.

Local capacity

In the United States, the management, financial, or technical ability of local communities to confront public problems effectively at the local level. See also CAPACITY BUILDING.

喜好。[参见"住房选择"]

居住面积　Jūzhù miànjī

在中国，指居住单元中用于起居、就寝房间的净面积。在城市规划中一般采用毛面积。[参见"居住面积密度"、"居住建筑面积"、"使用面积"]

借款条件　Jièkuǎn tiáojiàn

指借款合约中规定的偿还贷款的条件。

贷款与价值比　Dàikuǎn yǔ jiàzhíbǐ

在美国，指关于一项产业的贷款金额与该产业的估计价值之比。这种比值是影响贷款利率的因素之一，比值愈大，利率愈高。

地方能力　Dìfāng nénglì

在美国，指地方社区能有效地解决地方一级的公共问题所具备的管理、财政和技术能力。[参见"扩大能力"]

Local government

The authority charged with making public policy and with carrying out public administrative functions at the municipal or county level. In the United States, local government is the governmental authority immediately below state government. See also STATE GOVERNMENT.

Local project

In China, a construction project of an enterprise or institution that is administered by local government at the provincial, county, and municipal level. See also NATIONAL PROJECT.

Location Theory

The body of economic theory which seeks to explain the forces and factors that determine the location of economic activity, especially spatial patterns of companies, agriculture, towns, and households within towns. Central Place Theory and Concentric Zone Theory are part of Location Theory. See also CENTRAL PLACE THEORY, CONCENTRIC ZONE THEORY, and URBAN ECONOMICS.

Loop street

In the United States, a street that forms a circular pattern and either connects a larger street at two

地方政府 Dìfāng zhèngfǔ

负责制定政策并行使公共管理职能的市或县一级权力机构。在美国，地方政府直属州政府。[参见"州政府"]

地方项目 Dìfāng xiàngmù

在中国，指属于地方的管理建设项目，即由省、市、县直接管理的行政、企业、事业单位的建设项目。[参见"国家项目"]

位置理论 Wèizhì lǐlùn

试图解释决定经济活动分布状况的动力和因素的一种经济理论，尤其是关于公司企业、农业、市镇及市镇内居民的空间分布形态等。中心地带理论和同心区理论是位置理论的一部分。[参见"中心地带理论"、"同心区理论"、"城市经济学"]

半环路 Bànhuánlù

在美国，指一种环状道路，两端与大路相连或自身封闭，是进入居住区内或活动中心的通道。

points or closes upon itself, serving as an access route to a residential area or activity center.

Lot line

The official boundary of a measured land parcel.

Low-and-moderate-income housing

See ASSISTED HOUSING.

Low-rise building

See BUILDING HEIGHT.

Low-rise/high density

An approach to residential housing development that involves constructing a relatively large number of residential buildings a few stories high on the development site with relatively little space between them. See also HIGH-RISE/HIGH DENSITY and HIGH-RISE/LOW DENSITY.

地界 Dìjiè

经过丈量的地块的正式界线。

中低收入住房 Zhōngdī shōurù zhùfáng

[见"公助住房"]

低层建筑 Dīcéng jiànzhù

[见"建筑高度"]

低层高密度 Dīcéng gāomìdù

住房建设的一种方法,即在建设用地内建造比较多的低层住宅,而留出比较少的室外空地。[参见"高层高密度"、"高层低密度"]

M

Main street

A street that leads from the arterial road in an urban area, or a major street in a small town. In the United States, small towns typically have a street called "Main Street" that is the center of commercial activity.

Maintenance bond

In the United States, a document, given by the contractor to the owner, guaranteeing to rectify defects in workmanship or materials for a specified time following completion of the project. A one-year maintenance bond is normally included in the performance bond. See also BOND and PERFORMANCE BOND.

Management structure

The management of a typical entity falls into a hierarchy with three levels. Executive or top management of private organizations includes owners, corporate heads, and executive officers who make the policy decisions for the organization. Middle management encompasses contract managers, project managers, and project engineers who are responsible for the administrative duties associated with project work, e.g., scheduling, regulation of work flow, and resource allocation. At the lowest

主要街道 Zhǔyào jiēdào

城市地区内与主干道相连接的街道或小城镇中的大街。在美国,小城镇中都有一条称为"主街"的街道,是商业活动的中心。

维修保证书 Wéixiū bǎozhèngshū

在美国,由承包人向业主开具的保证书,保证在工程完工以后的一定期限内对因施工或材料质量所造成的损坏负责修理。履行合同保单内通常附有为期一年的维修保证书。[参见"保证书"、"履行合同保单"]

管理结构 Guǎnlǐ jiégòu

一种典型的分三个管理层次的机构体制。私营组织的行政管理或上层管理人员包括:业主、公司领导和行政主管人员,是该组织的决策层。中层管理人员包括:合同经理、项目经理和项目工程师,负责与工程项目有关的行政管理,例如制订计划、调节工作流程、分配资源等。基层管理人员即施工管理人员包括:工段长、工长、工地监督、负责指挥施工人员、执行与施工过程直接有关的管理工作。

level of supervision, production management, including superintendents, foremen, and supervisors, directs the work of employees engaged in production and performs administrative tasks directly related to the work process.

Manufactured housing

1. In the U.S. building industry, the term refers generally to modular housing, panelized houses, mobile homes, and any other type of housing whose components are predominantly factory-produced. See also INDUSTRIALIZED BUILDING SYSTEM, MODULAR HOUSING, and PANELIZED HOUSING. 2. The term in the specific sense defined in the National Manufactured Construction and Standards Act of 1974 designates a one-story movable living unit on a chassis with wheels, with a living area of at least 320 square feet, completed at the factory, usable as permanent housing, and equipped with means for connecting water, sanitary, and electric facilities. The unit is popularly called a mobile home even though the structure rarely moves after it has been placed on site. This is the least expensive housing in the United States and the only form that can qualify as "low cost" housing. Manufactured housing is the only type of housing in the United States constructed to meet a federally administered building code.

工厂预制住房 Gōngchǎng yùzhì zhùfáng

1. 在美国建筑工业中，通常指盒子结构式房屋、大板装配式房屋、活动住房以及主要构件由工厂预制的其他各类住房。[参见"工业化建筑体系"、"模数制住房"、"大板式住房"]

2. 按美国1974年颁布的《国家工厂制造的房屋及标准法》中特定的含义，指底部带轮的单层活动住房。其居住面积至少为320平方英尺。这种房屋，完全在工厂制造，有接通电和上下水的连接装置，可作永久性住房用。通常称为活动住房，但安装后很少有移动的。是美国最廉价的住房，是唯一名副其实的"低造价"住房，也是美国唯一按联邦建筑规范所建的住房。

Marginal land

A land tract of such poor quality that it scarcely warrants the expense of farming and is unsuitable for urban use without extensive site preparation costs.

Market analysis

Research conducted to determine the characteristics and number of potential purchasers of a product or products within a given area, e.g., housing market analysis. See also MARKET AREA.

Market area

The region in which potential or current users of a product or service are located. See also MARKET ANALYSIS.

Market price

In economic theory, the property or product price at which informed buyers and sellers are willing but not compelled to do business, as determined by supply and demand factors on the open marketplace. The price represents the market value of the product or property, If the property has been on the market for a reasonable time and payment is made at terms usual for the location and type of property. See also MARKETPLACE.

边角地 Biānjiǎodì

指土质差且不适于耕种的土地或不耗资处理就不能用于城市建设的土地。

市场分析 Shìchǎng fēnxī

对于某种产品的特性及其在一定地区内可能有多少买主所做的分析研究,如住房市场分析。[参见"市场所在地"]

市场所在地 Shìchǎng suǒzàidì

指某项产品或服务当前或潜在顾客所在的地方。[参见"市场分析"]

市场价格 Shìchǎng jiàgé

在经济理论中,由公开市场的供求因素所决定,为消息灵通的买方和卖方愿意据以进行交易的产业或产品的价格,即为市场价格。这种价格代表产品的市场价值,对于产业来说,如果一项产业进入市场已经有了相当长的时间,并且是按该产业坐落位置和类型的通常付款条件付款成交的,则该产业的市场价格即为其市场价值。[参见"集市"]

Market rate loan

In the United States, a loan granted by a lending, institution to a borrower at the prevailing rate of interest current at a given time in lending market dealings.

Market rent

In the United States, the amount a tenant pays a landlord for the use of property, as determined by current prices for comparable property in the absence of rent controls or other policies that distort supply and demand relationships. Under HUD guidelines for rent subsidies, the fair market rent is the gross rent allowable for a unit of a specified size within a Metropolitan Statistical Area. See also HOUSING ALLOWANCE and RENTAL HOUSING.

Marketplace

1. An open area in the center of a village, town, or city to which goods are brought and where they are sold. In the United States, marketplaces have been replaced by farmers' markets and flea markets held at regular or irregular intervals on squares or parking areas or in large halls.

2. In open market economic systems, an abstract concept designating the sphere of economic activity in which supply and demand relationships, as well

市场利率贷款 Shìchǎng lìlǜ dàikuǎn

在美国,指由贷款组织按借贷市场上一定时间内的通行利率贷给借款人的贷款。

市场租价 Shìchǎng zūjià

在美国,当不存在租金管制或其他使供求关系失常的政策的情况下,为使用一项房地产而由租户按现时租价付给房产主的租金。按美国住房与城市发展部关于租金津贴的规定,所谓公平的市场租价,就是在某一大都市统计区内一套一定大小的住房的允许总租价。[参见"住房津贴"、"租用住房"]

集市 Jíshì

1. 城市或村镇中作为贸易场所的空旷场地。美国的集市已为农民市场或"跳蚤"市场所代替。这种市场一般都设在广场、停车场或大型公共会场中,以定期或不定期的形式进行交易。

2. 在开放型市场经济体制中,亦可用作抽象概念。指通过私人交易确立供求关系及物价的经济活动。[参见"市场价格"]

as prices, are determined through private trading and sales transactions. See also MARKET PRICE.

Mass transit

Public systems for transporting large numbers of passengers rapidly and conveniently within urban areas, generally by bus, subway, or rail. Also called rapid transit.

Master plan

In the United States, a comprehensive document showing the general long-range physical design proposed for development of an entire community, or area, including maps, illustrations, and tables. The plan is the result of the formal process of comprehensive planning and may also be called a comprehensive plan. See also COMPREHENSIVE PLANNING.

Matching grant

In the United States, the sum of money allocated for projects to states or localities by the federal government on the condition that the states or localities dedicate sufficient local revenues to the program to satisfy the legally required matching ratio. The Interstate Highway System was built using matching grants. See also GRANT.

大量客运　Dàliàng kèyùn

在城市地区快速便捷大量运送乘客的公共运输系统。通常指公共汽车、地铁或火车，又称"快速客运"。

规划总图　Guīhuà zǒngtú

在美国，指为整个社区或地区开发所做的远期实体设计的综合性文件，包括地图、图形和表格。规划总图是正规化的总体规划过程的产物，又可称为"总体规划图"。[参见"总体规划"]

相应赠款　Xiāngyìng zèngkuǎn

在美国，指联邦政府为兴建工程项目而给州或地方政府的津贴，接受津贴的条件是州或地方政府按法律规定的比例，拿出足够的税收收益投资于该项工程。如某些州际高速公路即采用这种办法建成。[参见"赠款"]

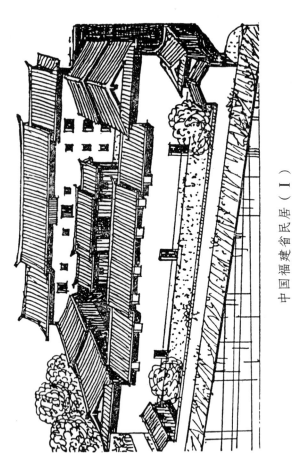

中国福建省民居（Ⅰ）
Vernacular dwelling in Fujian Province (Ⅰ)

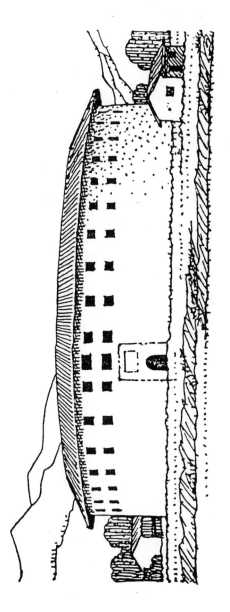

中国福建省民居（Ⅱ）
Vernacular dwelling in Fujian Province (Ⅱ)

Material requirements per square meter

In China, the average amount of materials needed to construct a square meter of floor area on a construction project. See also NORM FOR ESTIMATING MATERIAL REQUIREMENTS.

Maximum occupancy

The greatest permissible number of persons that can be housed in a particular dwelling. In the United States, the maximum is usually determined on the basis of health and safety considerations, e.g., fire regulations. See also OCCUPANCY.

Medium-sized city

See CITY SIZE.

Metropolis

See CITY SIZE.

Metropolitan area

A heavily populated area that includes a large city or metropolis and the surrounding towns and suburbs that are subject to its direct influence, have common interests, and share economic and social activities.

Metropolitan Statistical Area (MSA)

As established for statistical purposes by the U.S.

单方用料 Dānfāng yòngliào

在中国，建设工程中每平方米建筑面积所耗用的材料的平均数量。[参见"材料定额"]

最多居住人数 Zuìduō jūzhù rénshù

一套住房中允许居住的最多人数。在美国，一般根据卫生、安全（如防火规范）等方面的考虑加以规定。[参见"占用"]

中等城市 Zhōngděng chéngshì

[见"城市规模"]

大都市 Dàdūshì

[见"城市规模"]

大都市地区 Dàdūshì dìqū

人口众多的地区，包括一个大城市或大都市及在其周围，受其直接影响，相互间有共同利益并从事共同经济与社会活动的城镇和郊区。

大都市统计区 Dàdūshì tǒngjìqū

根据国家统计需要，美国管理及预算署和美国人

Office of Management and Budget and the U.S. Bureau of the Census, a geographic area, usually a county or contiguous group of counties, that contains a central city with at least 50,000 residents, including the surrounding suburban areas or twin cities with a population of more than 50,000 inhabitants. The name was changed in 1983 from Standard Metropolitan Statistical Area (SMSA). See also CENTRAL CITY.

Microclimate

The prevailing temperature, humidity, and atmospheric conditions in a local area that may vary in size from a crevice to a fairly large section of land as long as the conditions within the area remain relatively uniform. Microclimate is often a factor in land use planning.

Middle management

See MANAGEMENT STRUCTURE.

Mid-rise building

See BUILDING HEIGHT.

Mixed economy

An economic system with both public and private ownership of major enterprises. In such an economy, both sectors are strong and affect each other in

口普查局共同划定的包括一个或几个县的地理区域单位，其中有一个至少拥有 50000 人口的中心城市及其周围郊区，或人口超过 50000 人的两个城市。1983 年前称为"标准大都市统计区"。［参见"中心城市"］

小气候　Xiǎoqìhòu

指范围大小不等的局部地区内相对一致的温度、湿度和气候条件。小气候常是土地利用规划必须考虑的因素之一。

中层管理　Zhōngcéng guǎnlǐ

［见"管理结构"］

中层建筑　Zhōngcéng jiànzhù

［见"建筑高度"］

混合经济　Hùnhé jīngjì

一种包括公营和私营大企业的经济体系。在这种体系中，双方均财力雄厚并相互影响。

substantial ways.

Mixed housing development

A large-scale housing project that includes both private and publicly assisted dwellings. Such developments are not common in the United States.

Mixed team

See CONSTRUCTION TEAM.

Mixed-use development (MXD)

In the United States, land use projects that are carefully coordinated according to a master plan. Mixed-use developments vary in size and are characterized by multifunction buildings (e.g., shopping malls, hotels, and apartments), a compact configuration of components, and intensive use of urban land. Such developments frequently revitalize deteriorating downtown areas of large cities. See also MIXED-USE ZONING.

Mixed-use zone

In China, an urban (literally "complex") zone containing both the residential quarters and the work places of inhabitants, as well as various service facilities. The use pattern may be planned or occur spontaneously.

住房联合开发 Zhùfáng liánhé kāifā

由私人和公家共同资助的大规模住房建设项目。但这种开发方式在美国并不普遍。

混合施工队 Hùnhé shīgōngduì

［见"施工队"］

多用途建设 Duōyòngtú jiànshè

在美国，指根据规划总图精心协调的一地多用计划。多用途建设的规模大小各异；以建造多功能综合楼为其特征（如大型购物商场、旅馆和公寓等）。各组成部分之间结构紧密，城市土地利用率高，此种建设形式往往用来复兴大城市中衰落的闹市区。［参见"多用途区划"］

多用途区 Duōyòngtúqū

在中国，指一种经过规划或自然形成的综合性城市地区，其中包括居住区、为区内居民提供就业的场所以及各种服务设施。

Mixed-use zoning

In the United States, zoning that permits use of land for a variety of purposes. All land uses may be allowed and performance standards, prescribed for size, location, and density, or permitted uses may be specifically enumerated. See also MIXED-USE DEVELOPMENT, PLANNED UNIT DEVELOPMENT, and ZONING.

Mobile home

See MANUFACTURED HOUSING.

Mobility

Capacity or facility for movement or relocation, as in the case of population or the labor force. The U.S. Census Bureau obtains data on mobility status by asking subjects of population surveys for the addresses of their residences twelve months earlier or five years earlier, depending on the particular survey. Statistics indicate that about one-fifth of all American families relocate every year.

Modal split

The breakdown of overall travel or traffic figures by type of transportation, e.g., commuter traffic consisting of 65 percent automobile transportation, 10 percent bus transportation, etc.

多用途区划 Duōyòngtú qūhuà

在美国,指允许有不同用途的土地区划。土地的用途不受限制,仅在功能标准中对不同用途土地的大小、位置和密度做出规定或者具体列出各种允许用途。[参见"多用途建设"、"有规划的地段建设"、"区划"]

活动住房 Huódòng zhùfáng

[见"工厂预制住房"]

人口流动 Rénkǒu liúdòng

指人口或劳动力迁移时的移动或迁居的能力条件。美国人口普查局根据不同调查需要,请调查对象提供他们12个月或五年以前的地址,从而取得人口流动状况的数据。统计表明,有1/5的美国家庭每年搬一次家。

交通工具分类 Jiāotōng gōngjù fēnlèi

按运输方式对运行和交通的统计数字进行分类。如通勤交通中包括65%的小汽车运输、10%的公共汽车运输等。

Model code

An exemplary set of regulations designed to serve as a uniform standard, e.g., a model building code. See also BUILDING CODES AND STANDARDS.

Modular coordination

Planning and design of housing to accommodate standard dimensions of common building materials so that scrap and waste are minimized, and buildings are erected more quickly, efficiently, and cost-effectively than they would be otherwise. See also DIMENSIONAL COORDINATION, INDUSTRIALIZED BUILDING SYSTEM, and MODULE.

Modular housing

Three-dimensional, factory-produced units shipped to a site fully erected and placed on a finished foundation. The modular housing unit must conform to local building code requirements. In the U.S. building industry, modular housing is considered a type of manufactured housing. See also INDUSTRIALIZED BUILDING SYSTEM and MANUFACTURED HOUSING.

Module

A unit of standard size used in design and construction to facilitate assembly of building parts and to reduce

样板法规 Yàngbǎn fǎguī

为统一标准而制定的一套示范性规程,例如样板建筑法规。[参见"建筑法规与标准"]

模数协调 Móshù xiétiáo

适应一般建筑材料尺寸的住宅规划与设计,以尽量减少材料的浪费,从而能更迅速、有效、经济地建造住房。[参见"尺寸协调"、"工业化建筑体系"、"模数"]

模数制住房 Móshùzhì zhùfáng

由工厂预制的房屋单元,运到现场即可在事先做好的基础上安装就位。在美国建筑业中,模数制住房是工厂预制住房之一种,必须符合当地建筑法规的要求。[参见"工业化建筑体系"、"工厂预制住房"]

模数 Móshù

为便于组装房屋部件并减少材料浪费而在设计和施工中采用的一种标准尺寸单位。[参见"工业化

waste of materials. See also INDUSTRIALIZED BUILDING SYSTEM, MANUFACTURED HOUSING, and MODULAR COORDINATION.

Monitoring

Checking the progress of a project or program, conformance to standards and codes, construction quality, and so forth. See also QUALITY CONTROL.

Mortgage

In the United States, a borrowing arrangement in which the borrower (mortgagor) gives the lender (mortgagee) a lien on property as security for the payment of an obligation. The borrower continues to use the property, and when the obligation is fully settled, the lien is removed. See also ADJUSTABLE RATE MORTGAGE, AMORTIZATION, BALLOON MORTGAGE, DELINQUENT MORTGAGE, FIXED-RATE MORTGAGE, GRADUATED PAYMENT MORTGAGE, LIEN, MORTGAGE BANK, MORTGAGE REVENUE BOND, PRIMARY MORTGAGE MARKET, SECONDARY MORTGAGE MARKET, SHARED APPRECIATION MORTGAGE, and UNDERWRITING.

Mortgage bank

In the United States, a financial institution that

建筑系统"、"工厂预制住房"、"模数协调"］

施工监督　Shīgōng jiāndū

对于工程或计划的进度，以及是否符合标准、规范和施工质量等的检查。［参见"质量控制"］

抵押贷款　Dǐyā dàikuǎn

美国的一种贷款活动。借款人（抵押人）以一项财产（通常为不动产）的扣押权授予贷款人（受押人），作为偿还贷款的保证。借款人可继续使用该项财产，当贷款全部还清时，产业的扣押权即失效。［参见"可调利率抵押贷款"、"分期偿还"、"特大尾数抵押贷款"、"拖欠抵押贷款"、"固定利率抵押贷款"、"递增偿还抵押贷款"、"留置权"、"抵押贷款银行"、"抵押收益债券"、"初级抵押贷款市场"、"二级抵押贷款市场"、"分享增值抵押贷款"、"风险担保"］

抵押贷款银行　Dǐyā dàikuǎn yínháng

美国的一种专门从事抵押贷款的金融机构，从事收

specializes in mortgage financing, charging set fees for such functions as collecting payments, maintaining mortgage, records, and serving as escrow agents. The mortgage bank also provides the link between the demands of home buyers in the primary mortgage market and the investment funds available through the secondary mortgage market. See also MORTGAGE, PRIMARY MORTGAGE MARKET, and SECONDARY MORTGAGE MARKET.

Mortgage revenue bond

In the United States, a bond issued by a local government to finance private mortgages at below-market interest rates to encourage home building in the area. Such bonds are backed by the revenues from the mortgages and are tax-exempt. See also BOND and MORTGAGE.

MSA

See METROPOLITAN STATISTICAL AREA.

Multifamily housing

Residential structures designed to house more than one family. Multifamily structures have only one exterior entrance for a number of units, as opposed to townhouses, which have separate exterior entrances for each unit. See also APARTMENT HOUSE.

款、保存抵押契据、作为第三方抵押品保存人等业务，并为此收取业务费。抵押贷款银行也是沟通购房者在初级抵押贷款市场中的需要和二级抵押贷款市场中的投资资金的渠道。[参见"抵押贷款"、"初级抵押贷款市场"、"二级抵押贷款市场"]

抵押收益债券 Dǐyā shōuyì zhàiquàn

在美国，由地方政府发行的一种债券，用低于市面的利率资助私人抵押贷款，以鼓励在本地区内建造住房。这种债券以抵押收益为后盾，并可免税。[参见"债券"、"抵押贷款"]

大都市统计区（缩写） Dàdūshì tǒngjìqū (suōxiě)

[见"大都市统计区"]

多户住房 Duōhù zhùfáng

为多户居住所设计的居住建筑，由一个出入口通向各居住单元。不同于各单元有单独出入口的市镇联立式住宅。[参见"公寓楼"]

Multiplier effect

In economic theory, an increase in overall income, and consequently in spending and savings, resulting from injection of funds into a local economy. Expenditures for a program in a particular sector, e.g., construction, most affect the income of the workers immediately involved, but through the resultant increases in spending and savings, the increases in income are spread in diminishing amounts to workers in other sectors of the economy. See also LEVERAGING.

Multistory factory

An industrial building with separate floors leased to different firms engaging in the same or different industries. In British jurisdictions, called a flatted factory.

Municipal housing management bureau

Organization in China at the municipal level that is administered by the municipal government and is in charge of the housing stock and public or industrial buildings not owned by institutions. The housing management hierarchy includes the municipal housing management bureau, the district housing management bureau, and the housing management office. The district bureau is under the concurrent

收入增殖作用 Shōurù zēngzhí zuòyòng

在经济理论中,指由于将资金注入某局部经济而引起收入的全面增长以及相应的消费和储蓄的增长。对某一部门例如工程建设的一项计划进行投资,将对与该部门直接有关的工人的收入产生最大影响,而通过由此引起的消费和储蓄的增长,还能使收入的增长扩及其他部门,使其他经济部门工人的收入相应增长,但其增长量则随各部门与核心部门相关程度的减弱而递减。[参见"杠杆作用"]

分层出租厂房 Fēncéng chūzū chǎngfáng

一种供出租用的多层厂房,各层可分别由不同的工厂租用。这些工厂可属于相同或不同的工业。

市房管局 Shì fángguǎnjú

中国市级主管住房及其他建筑物(不包括属于企、事业单位所有者)的机构,受市政府领导,其建制包括市房管局、区房管局和房管所三级。区房管局受区政府及市房管局的双重领导,房管所受街道办事处及区房管局的双重领导。[参见"住房申请权"、"住房分配"]

administration of the district government and the municipal housing management bureau; the housing management office is under the concurrent administration of the subdistrict office and the district housing management bureau. See also ENTITLEMENT TO HOUSING and OBLIGATION OF ORGANIZATIONS TO PROVIDE HOUSING.

Municipality

1. In China, a city empowered by the State Council to maintain its own government and organizational system, with jurisdiction within a designated region. Certain major cities, i.e., Beijing, Shanghai, and Tianjin, are centrally administered municipalities that report directly to the central government. Provincially administered municipalities, including provincial capitals, are under the authority of the provincial government. See also AUTONOMOUS REGION.

2. In the United States, the term generally applies to a local government, regardless of size, which is chartered by the state to govern itself.

Municipally administered county

See COUNTY.

Municipally affiliated county

See COUNTY.

市 Shì

1.在中国,指经国务院批准设市建制,并划定其管辖地区的城市。市分直辖市和省辖市两种,特大城市如北京、上海和天津由中央政府直接管辖。省辖市包括省会,由省、自治区地方政府直接管辖。[参见"自治区"]

2.在美国,指由州政府确定,具有行政权力的地方政府。其规模大小各有不同。

市辖县 Shìxiáxiàn

[见"县"]

市管县 Shìguǎnxiàn

[见"县"]

Municipally owned housing

In China, housing owned by the housing management office of the city government, to be allotted to city residents. See also ENTITLEMENT TO HOUSING and OBLIGATION OF ORGANIZATIONS TO PROVIDE HOUSING.

Mutual housing

An association of cooperatives united through a Mutual Board to insure financially sound management practices for the existing housing and to raise funds for new construction. Each member co-op is responsible for its own property maintenance and is represented on the Mutual Board. To encourage future building programs, representatives of various outside public interest organizations, e.g., churches or private, non-profit social agencies, are also Board members. There are about twelve different types of mutual housing associations in the United States, many of them based on Swedish or German models. See also HOUSING COOPERATIVE.

Mutual savings bank

In the United States, a state-chartered banking organization without capital stock that receives deposits and distributes legally limited earnings to depositors. Such banks are geared to small savers,

房管部门住房 Fángguǎnbùmén zhùfáng

在中国,指产权属于城市房屋管理部门并由其分配给本市居民的住房。[参见"住房申请权"、"住房分配"]

互助住房协会 Hùzhù zhùfáng xiéhuì

以互助委员会形式组织起来的一种住房合作社联合协会,旨在保证现有住房的财务管理,并为新建住房筹措资金。每个加入该组织的住房合作社都有代表参加互助委员会。房产的维修则分别由各合作社自行承担。为了鼓励兴建住房,各公共组织如教会、私人机构、非营利性社会机构的代表都可参加互助委员会。在美国,约有12种不同形式的互助住房协会,大多参照瑞典或德国模式。[参见"住房合作社"]

互助储蓄银行 Hùzhù chǔxù yínháng

在美国,指由州政府设立的没有股本的金融机构。这种银行面向小额存户,接受他们的存款,同时将法定限额的收益分发给这些存户,使他们能将储蓄投资于抵押贷款和证券。[参见"初级抵押贷

permitting them to invest savings in mortgages and securities. See also PRIMARY MORTGAGE MARKET.

MXD

See MIXED-USE DEVELOPMENT.

款市场"]

多用途建设(缩写) Duōyòngtú jiànshè (suōxiě)
[见"多用途建设"]

N

National economic and social development planning

In China at present, macro-economic planning for the national economy and social development, with goals to be achieved within a period of time, usually five years. The program encompasses the major aspects of production, distribution, and consumption, including planning of the national income, agricultural and industrial production, urban and rural development and the building industry, trade and finance, science and technology, social welfare, population growth, environmental protection, cultural activities, and so forth. See also LABOR EQUILIBRIUM.

National project

In China, a construction project under the supervision of the central government, i.e., a construction project of a government agency, enterprise, or institute under the direct leadership and control of a ministry of the State Council. See also LOCAL PROJECT.

Natural Conservation Zone

In China, a natural region declared by the State Council to possess such great ecological, resource, or scientific value that it warrants protection. An example is the Wolong Giant Panda Conservation Zone of Sichuan Province.

国民经济和社会发展计划 Guómín jīngjì hé shèhuì fāzhǎn jìhuà

在中国，指国民经济和社会发展的宏观计划，规定在一定时期（一般为五年）内所需达到的目标。计划包括生产、流通、分配、消费等主要方面，如国民收入、工农业生产、建筑业、城乡建设、环境保护、科学技术、商业与金融、人民生活和社会福利、文教卫生体育及人口等。[参见"劳动平衡"]

国家项目（部直属项目） Guójiā xiàngmù（bùzhíshǔ xiàngmù）

在中国，指属于中央管理的建设项目，即由国务院各部直接领导和管理的行政、企业、事业单位的建设项目。[参见"地方项目"]

自然保护区 Zìrán bǎohùqū

在中国，由国务院确定并批准的具有生态、资源和研究价值并需要予以保护的自然区域，如四川省卧龙大熊猫保护区等。

Natural population growth

Increase in the size of population as the result of births, as contrasted with growth through migration. See also POPULATION GROWTH FROM MIGRATION.

Neighborhood

In the United States, an unofficial social unit of residents living closely together in a relatively small area of a city or town, sharing to some degree common cultural, social, or economic interests, and possessing local facilities and institutions such as libraries and schools. See also SUBDIVISION.

Neighborhood committee

In China, an organization of inhabitants in a living quarter. The committee is responsible for such tasks as maintenance of sanitary conditions, security, settlement of disputes, learning programs, family planning, and juvenile after-school activities. See also HOUSING GROUP and NEIGHBORHOOD GROUP.

Neighborhood conservation

In the United States, efforts to preserve the buildings, local service facilities, and community atmosphere in older urban neighborhoods threatened by economic,

人口自然增长 Rénkǒu zìrán zēngzhǎng

因生育而增加的人口，区别于因迁移形成的人口增长。[参见"人口机械增长"]

邻里 Línlǐ

在美国，指一种非官方的社会单位，由市镇小片地区内住得较近的居民组成。他们在一定程度上有共同的文化、社会或经济利益，并共享一些地方设施与机构，如图书馆和学校等。[参见"分块土地"]

居民委员会 Jūmín wěiyuánhuì

中国城市居住区中的居民组织，负责组织居民清扫环境、治安、调解纠纷、学习、宣传计划生育及青少年的课余活动等。[参见"住宅组团"、"居民小组"]

邻里保护 Línlǐ bǎohù

在美国，指在经济、社会和物质上渐趋衰落，或旧建筑有危险被拆掉而代之以新建筑的城市旧邻里中，为保存原有建筑物、地区服务设施及社区

social, and physical deterioration or replacement by modern buildings. See also HISTORIC DISTRICT, HISTORIC PRESERVATION, and NEIGHBORHOOD REVITALIZATION.

Neighborhood group

In China, a grass roots unit which conducts public service activities arranged by the neighborhood committee. See also NEIGHBORHOOD COMMITTEE.

Neighborhood industry

In China, a small business usually run collectively by the subdistrict office in an urban area. See also COTTAGE INDUSTRY.

Neighborhood participation

See CITIZEN PARTICIPATION.

Neighborhood revitalization

In the United States, efforts to restore housing and to stimulate renewed interest in community social and commercial activity within city neighborhoods that have deteriorated. See also ADAPTIVE RE-USE, NEIGHBORHOOD CONSERVATION, URBAN REDEVELOPMENT, and URBAN REVITALIZATION.

Neighborhood watch

In the United States, an informal voluntary organization

气氛等所采取的措施。[参见"历史性市区"、"历史性建筑保护"、"邻里复苏"]

居民小组 Jūmín xiǎozǔ

中国居民的基层组织,受居委会领导,进行居委会所组织的有关公共活动。[参见"居民委员会"]

街道工厂 Jiēdào gōngchǎng

中国城市中,由街道办事处办的集体所有制工厂。[参见"家庭工业"]

邻里参与 Línlǐ cānyù

[见"市民参与"]

邻里复苏 Línlǐ fùsū

在美国,指在衰落的城市邻里中为修复住房和恢复社区中社会与商业活动的生气而采取的措施。[参见"适应性再使用"、"邻里保护"、"城市改造"、"城市复苏"]

邻里自卫 Línlǐ zìwèi

在美国,邻里中居民为共同抵御盗贼和其他犯罪

of neighbors, interested in mutual protection against burglary and other crimes. Group members, who are unarmed, stay on the alert for suspicious incidents, and contact police by radio or telephone if such incidents occur. Also called block watch. For a Chinese organization with certain analogous functions, see also VOLUNTARY SECURITY SERVICE.

Net leasable area

See LEASABLE AREA.

Net residential area

The actual area covered by housing measured over a number of sites, including adjacent yards and outbuildings.

Net residential density

See RESIDENTIAL DENSITY.

Net site area

In general, the area of a housing site measured between its property boundaries. Whether or not the measure includes all or a portion of the street area within or bounding the site must be determined for the specific case.

Network analysis

Definition of the components of an overall urban

活动而组成的非正式的志愿自卫组织,其成员不带武器,只对可疑行踪保持警惕。一旦发现问题,即用无线电或电话与警察联系,又称街区自卫。[有关中国的类似组织,参见"居民联防"]

可出租净面积 Kěchūzū jìngmiànjī

[见"可出租面积"]

居住用地净面积 Jūzhùyòngdì jìngmiànjī

居住区内若干住房基地净面积的总和,包括宅间院子及宅旁小建筑物用地。

居住净密度 Jūzhù jìngmìdù

[见"居住密度"]

住房基地净面积 Zhùfáng jīdì jìngmiànjī

通常指产业界线之内的住房基地面积。居住区内部道路的面积以及居住区四周道路的面积是否计算在内,则因情况而异。

网络分析 Wǎngluò fēnxī

确定整个城市交通系统的各构成部分,并对交通

transportation system and examination of the network's efficiency, cost, and effects on urban functions. See also TRAFFIC FORECAST.

New town

A large, newly built urban entity which has been developed according to a single master plan and is under the control or ownership of a single development entity. The new town has reasonably balanced land uses, creating a new community or major addition to an existing community, which includes most, if not all, of the basic services, and facilities normally associated with a city or town. The concept was developed in Great Britain and then adopted in the United States. Also called a new community. See also FREESTANDING NEW TOWN and SATELLITE CITY.

NGO

See NONGOVERNMENTAL ORGANIZATION.

Noise controls

Regulatory measures or guidelines instituted by government to reduce noise levels considered unacceptably high because of potential disturbance or injury to individuals, businesses, or public service organizations.

网的效率、成本及其对城市功能的影响所做的分析。[参见"交通预测"]

新城 Xīnchéng

根据一个规划总图建设起来的、土地使用合理平衡的大型新建城市实体。其所有权属一个开发单位或由其经营管理。新城的建设形成一个新的社区,或在原有社区的基础上增加新的内容,包括城市生活所需要的基本服务及设施。此概念起源于英国,后为美国采用。亦称"新社区"。[参见"独立新城"、"卫星城"]

非政府组织(缩写) Fēizhèngfǔ zǔzhī(suōxiě)

[见"非政府组织"]

噪声控制 Zàoshēng kòngzhì

政府防止过高的噪声对人、商业或公共服务机构的干扰或伤害而制定的条例或法规。

Nonconforming use

In the United States, use of land or a type of building not permissible under zoning ordinances. Nonconforming use results from adoption of new or revised ordinances that apply to both existing and new development; some types of nonconformance are allowed to continue, for example, building nonconformance of a physical character, while a schedule for amortization is defined in cases of extreme nonconformance. See also AMORTIZATION.

Nonconforming work

Construction project work that does not satisfy the requirements defined in the contract documents for the project. See also PUNCH LIST and SITE INSPECTION.

Nongovernmental organization (NGO)

In the United States, organized citizens' group without government affiliation that seeks to improve conditions in communities through active participation in the planning process.

Norm

In China, a quantitative index of the labor, material, or money required for the production activities of a construction enterprise to complete a certain

不符规定使用 Bùfú guīdìng shǐyòng

在美国,指土地利用或建筑类型违反区划法令的规定。导致这种不符合规定使用的原因,在于采用了对原有和新建设都有约束力的新法令或修订法令,有些建筑物的实体特征不符合规定,但仍准许继续使用。严重不符合规定者需逐步淘汰。[参见"逐步淘汰"]

不符合同工程 Bùfú hétóng gōngchéng

未按合同文件规定的要求施工的工程项目。[参见"建筑缺陷清单"、"现场监督"]

非政府组织 Fēizhèngfǔ zǔzhī

在美国,指不隶属于政府的公众组织。其目的在于通过积极参与规划过程,以改进社区条件。

定额 Dìng'é

在中国,指建筑安装企业在生产活动中为完成一定任务所消耗人力、物力、财力的数量标准。按其内容可分为费用定额、劳动定额、材料定额等。

construction task. These indices, some of which are statutory, are classified according to the component measured, such as service fee norms, norms or labor requirements, norms for material requirements, and so forth. See also CONSTRUCTION NORM, NORM FOR DETAILED ESTIMATES, NORM FOR ESTIMATING LABOR REQUIREMENTS, NORM FOR ESTIMATING MATERIAL REQUIREMENTS, NORM FOR PRELIMINARY ESTIMATES, SERVICE FEE NORM, and TEN-THOUSAND-YUAN NORM.

Norm for detailed estimates

In China, the normal amount of labor, material, or construction equipment (in machine-shifts, i.e., machine work measured in eight-hour shifts) required per unit measure, e.g., square meter or cubic meter, of various building components, such as foundation, or brickwork, as stipulated by State or regional authorities. Norms for detailed estimates are the basis for the preparation of detailed estimates and the settlement of accounts for completed projects, as well as for the drafting of construction management plans, for business accounting, and for calculation of project costs. See also DETAILED ESTIMATE and NORM.

有些定额由政府部门制定。[参见"施工定额"、"预算定额"、"劳动定额"、"材料定额"、"概算定额"、"取费标准"、"万元定额"]

预算定额 Yùsuàn dìng'é

在中国,指由国家或地区领导机关制定的建筑物各个组成部分按单位计量所需的人工、材料和施工机械台班数量的标准。例如基础工程所需的人工数、混凝土立方数和机械台班费。预算定额是编制设计预算、进行竣工结算的依据,也是编制施工组织设计、进行经济核算、考核工程成本的基础。[参见"设计预算"、"定额"]

Norm for estimating labor requirements

In China, the manhours required to produce certain product units or to complete certain units of work, as stipulated by the construction enterprise or higher authorities. See also LABOR REQUIREMENTS PER SQUARE METER and NORM.

Norm for estimating material requirements

In China, a norm that represents the normal amount of material required for a unit quantity of work. The norm may be for estimating material requirements to complete individual work elements or to construct a square meter of floor area for a particular type of building. See also MATERIAL REQUIREMENTS PER SQUARE METER, NORM, and TEN-THOUSAND-YUAN NORM.

Norm for preliminary estimates

In China, the cost of construction, the amount of labor, the amount of material, and the amount of construction equipment required for each unit of floor area, calculated on the basis of empirical data from completed buildings and structures and listed in the form of a table. These norms are the basis for preparation of preliminary estimates of construction costs and are also called preliminary estimate indices. See also NORM and PRELIMINARY ESTIMATE.

劳动定额 Láodòng dìng'é

在中国，指生产某一单位产品或完成某一单位工作所必须消耗的时间，由建筑安装企业或上级机关规定。[参见"单方用工"、"定额"]

材料定额 Cáiliào dìng'é

在中国，指生产某一单位产品或完成某一单位工作所必须消耗的材料数量，用以估计完成每部分工程所需材料或某一工程项目每平方米建筑面积所需的材料。[参见"单方用料"、"定额"、"万元定额"]

概算定额 Gàisuàn dìng'é

在中国，指单位建筑面积的造价以及劳动力、主要材料及主要施工机械的消耗量，是以整个建筑物或构筑物为单位的经验资料估算编定。概算定额是编制设计概算的依据，又称"概算指标"。[参见"定额"、"设计概算"]

Noxious industry

A type of industrial or manufacturing process or organization that produces hazardous wastes and pollutants likely to damage public health or quality of life. See also HAZARDOUS WASTES.

Nuclear family

A family which encompasses parents and their unmarried sons and daughters but does not include other relatives. See also EXTENDED FAMILY and FAMILY.

有害工业 Yǒuhài gōngyè

产生有害废物和污染物,足以危害公共卫生或生活环境质量的工业机构或制造过程。[参见"有害废物"]

核心家庭(小家庭) Héxīn jiātíng (Xiǎo jiātíng)

父母仅与其未婚子女同住一户的家庭。[参见"大家庭"、"家庭"]

O

O and D

See ORIGIN-DESTINATION SURVEY.

Obligation of organizations to provide housing

Under Chinese urban housing policy, all organizations must provide housing to their employees. Local housing management authorities are obligated to supply housing if the institutions are in no position to do so. See also EMPLOYEE DORMITORY, ENTITLEMENT TO HOUSING, FAMILY HOUSING, MUNICIPAL HOUSING MANAGEMENT BUREAU, MUNICIPALLY OWNED HOUSING, ORGANIZATION-OWNED HOUSING, and RIGHT OF OCCUPANCY.

Obsolescence

A loss in value of real property as the result of its becoming outdated. Obsolescence is regarded as functional if the design and construction of a structure no longer meet current housing needs and standards, and as economic if changes in the overall economic structure have eliminated the need for the building, e.g., railway stations in many parts of the United States.

Obsolescent housing

See OBSOLESCENCE.

起终点调查(缩写) Qǐzhōngdiǎn diàochá(suōxiě)

[见"起终点调查"]

住房分配 Zhùfáng fēnpèi

根据中国住房政策,单位有责任为其职工提供住房。如单位不能提供住房,应由城市房管部门负责提供。[参见"单身宿舍"、"住房申请权"、"家属宿舍"、"市房管局"、"房管部门住房"、"单位住房"、"居住权"]

过时 Guòshí

因年久而失去价值的房地产。有功能性和经济性两种情况。前者指一个建筑物的设计和建造不再能满足当前的住房需要及标准;后者指由于全面经济结构的改变,有些建筑物已不再需要,例如美国许多地方的火车站。

过时住房 Guòshí zhùfáng

[见"过时"]

Occupancy

1. The act of becoming a resident or the condition of being a resident of a property as a tenant or owner. Use by the owner of a portion of a project prior to completion is called partial occupancy. See also MAXIMUM OCCUPANCY, SHARED OCCUPANCY, and SINGLE OCCUPANCY.

2. A special use to which a property is put, for example, residential occupancy.

Occupancy permit

See CERTIFICATE OF OCCUPANCY.

Occupancy rate

The number of dwelling units inhabited by owner-occupants or tenants, as compared to the total number of existing units. See also VACANCY RATE.

Occupational structure

The types and numbers of jobs available in a particular country or region. The job outlook for the United States is reported at regular intervals in the *Occupational Outlook Handbook* published by the Bureau of Labor Statistics of the U.S. Department of Labor. See also STRUCTURAL UNEMPLOYMENT.

占用 Zhànyòng

1. 成为一处房地产的租用者或业主的行为或状况。在工程完工前,由业主使用部分房屋,称"部分占用"。[参见"最多居住人数"、"合住"、"独门独户"]

2. 房地产的一种特殊用途。例如,用于居住。

占用许可证 Zhànyòng xǔkězhèng

[见"符合使用证明书"]

住房占用率 Zhùfáng zhànyònglǜ

由房主或租户居住的住房单元数与住房单元总数之比。[参见"空房率"]

职业结构 Zhíyè jiégòu

指某一国家或区域内职业的类型和数量。由美国劳动部劳工统计局出版《职业展望手册》,定期报道美国职业结构情况。[参见"结构性失业"]

Only-child family

In China, a family in which a couple has agreed to have only one child, usually evidenced by an only-child certificate. See also FAMILY.

Open space

The portion of a housing development or urban area which is not built up and may be maintained in its natural state to promote conservation or aesthetic goals, or used for recreational purposes and facilities. Open space can also refer to any area of a site not covered by structures, including walkways, driveways, and parking space.

Open system

Components or materials manufactured according to standard specifications so that equivalent brands or types can be used interchangeably in the building process. See also CLOSED SYSTEM and DIMENSIONAL COORDINATION.

Operating budget

A detailed plan of expenditures and revenue estimates for current operations of a public or private organization over the course of a year.

Operation Breakthrough

A research and development (R&D) program of the

独生子女户 Dúshēng zǐnǚ hù

在中国，指一对夫妇只生一个孩子的家庭，通常以领得独生子女证为准。[参见"家庭"]

空地 Kòngdì

为保证环境或美观的目的，在住房建设或城市地区中留出一定不予建造的地段，或保持其自然状态，或用作游乐场地。空地也可指任何在其上没有建筑物的地方，例如人行道、车道和停车场等。

通用建筑体系 Tōngyòng jiànzhù tǐxì

构配件或材料按标准规范制作的一种建筑体系，在建设过程中可以用不同厂家生产的同类型构配件互换。[参见"专用建筑体系"、"尺寸协调"]

经营预算 Jīngyíng yùsuàn

公营或私营机构预计在年度范围内从事业务经营的收入和支出的详细计划。

突破行动 Tūpò xíngdòng

指美国住房与城市发展部 1969—1972 年的研究和

U.S. Department of Housing and Urban Development (HUD) from 1969 to 1972. On nine sites across the U.S., 2950 housing units were built following industrialized building methods. The program encouraged innovations in several areas: new building systems and products, site development, construction management, and performance-based building codes.

Operations research (OR)

Application of mathematical methods to the analysis of complex problems in activities of governmental, commercial, and military organizations.

Opportunity cost

Benefits foregone because an alternative to the current course of action was not chosen. For example, the opportunity cost for funds invested in housing may be the rate of return obtainable from the same funds invested in the stock market or in a bank.

Optimal land utilization

In China, the efficient use of land for capital construction, e.g., by taking up little or no farmland, by using sloped and nonarable lands for building, and by controlling building density in urban planning.

发展计划。在美国九个地方执行了此项计划，采用工业化建筑方法建成 2950 个住房单元。此计划鼓励在多方面进行革新：新的建筑体系和产品、建筑场地的开发、施工管理、采用功能法规等。

运筹学 Yùnchóuxué

运用数学方法分析政府、商业和军事机构中的复杂问题的一门学问。

机会成本 Jīhuì chéngběn

因从事目前活动而放弃另一种机会所可能带来的收益。例如，若以一笔资金投资于建造住房，其机会成本为以同一笔将资金投资于股票市场或存入银行所能获得的利息。

节约用地 Jiéyuē yòngdì

在中国，指基本建设中对于土地的节约。例如少占或不占农田，利用坡地和不宜种植的土地，以及在城市规划中对建筑物密度的控制。

Option

1. A course or alternative available at any given decision point.
2. In the United States, the right granted by contract to buy or sell securities or property at a specific price within a stipulated time period.

Options paper

A report, especially common in government, outlining alternatives for a policy decision, with a discussion of the advantages and disadvantages of each alternative.

OR

See OPERATIONS RESEARCH.

Ordinance

See BUILDING CODES AND STANDARDS.

Organization-owned housing

In China, housing owned by organizations, institutions, and enterprises to be allotted to their staff members and workers, as opposed to housing owned by the city government. See also OBLIGATION OF ORGANIZATIONS TO PROVIDE HOUSING.

1. 供选择方案 Gōngxuǎnzé fāng'àn
供决策时挑选的不同方案。

2. 买卖权 Mǎimàiquán
在美国，合同允许在规定期间以特定价格买、卖股票或房地产的权利。

决策资料 Juécè zīliào
通常由政府机构发表的报告，扼要介绍可供某项决策采用的各种替换方案的优缺点。

运筹学（缩写） Yùnchóuxué（sōuxiě）
［见"运筹学"］

法令 Fǎlìng
［见"建筑法规与标准"］

单位住房 Dānwèi zhùfáng
在中国，指产权属于机关、事业单位、企业单位并由其分配给本单位职工的住房，以有别于房管部门住房。［参见"住房分配"］

Organization-raised funds

In China, one of the financial resources for construction projects, referring to funds raised by the ministries of the State Council, provinces, municipalities, and counties, as well as by enterprises and institutes for their own use. See also INVESTMENT IN CAPITAL CONSTRUCTION.

Origin-destination survey (O and D)

A transportation study that records the numbers of trips of transportation vehicles and the beginning and ending points of the trips, usually by geographic zones.

Outer suburban district

See SUBURB.

Outskirts

See URBAN FRINGE.

Overall construction site plan

In China, a general layout plan of a construction site showing the locations of permanent structures already in existence and of those to be built, as well as construction site facilities and work areas. The plan is a major constituent part of a construction management plan. See also CONSTRUCTION

自筹资金 Zìchóu zījīn

中国建设项目资金来源之一,由国务院各部以及各省、市、县和企业、事业单位自己筹集的用于建设项目的资金。[参见"基本建设投资"]

起终点调查 Qǐzhōngdiǎn diàochá

在美国,指在一定地区内记录交通车辆行驶次数及其起终点的调查研究。

远郊区 Yuǎnjiāoqū

[见"郊区"]

城市外围 Chéngshì wàiwéi

[见"城市边缘"]

施工总平面图 Shīgōng zǒngpíngmiàntú

在中国,指施工现场的总体平面图,是施工组织设计的重要组成部分。图上除绘有已建和拟建的永久性建筑物外,还标明各种临时设施以及进行准备工作的场所。[参见"施工组织设计"、"临时设施"]

MANAGEMENT PLAN and CONSTRUCTION SITE FACILITIES.

Overcrowded household

See HOUSING SHORTAGE.

Overhead

The expenses for management, administration, facility maintenance, and utilities necessarily incurred to operate any business. See also INDIRECT EXPENSES.

Owner

The architect's or engineer's client with whom an agreement has been signed for a project. The project owner may be a government agency, an institution, a corporation, or an individual contractually obligated to pay for the work performed. See also CLIENT, CONTRACT, CONTRACTOR, OWNER-BUILDER, and OWNER-OCCUPANT.

Owner-builder

In the United States, a person acting as his/her own general contractor in building housing for his/her own use. See also OWNER and SELF-HELP HOUSING.

居住拥挤户　Jūzhù yōngjǐhù

［见"住房短缺"］

管理费　Guǎnlǐfèi

为经营任何业务所需的行政管理、设施维修以及公用事业等费用。［参见"间接费"］

业主　Yèzhǔ

建筑师或工程师的委托人，签订合同承担支付工程价款的工程所有者，可以是政府机构、事业单位、公司或个人。［参见"客户"、"合同"、"承包人"、"自建业主"、"自住业主"］

自建业主　Zìjiàn yèzhǔ

在美国，指自己担任总承包人建造自用住房的业主。［参见"业主"、"自建住房"］

Owner-occupant

In the United States, a person who both owns and occupies a commercial or residential property. See also OWNER.

自住业主 Zìzhù yèzhǔ

在美国,指住在产权归本人所有的商业或居住建筑中的人。[参见"业主"]

P

Package deal

In the United States, a business arrangement in which responsibility for all the functions in a process are assumed by one contractor, e.g., the design-build process, supervised by a competent package builder engaged to assist in all phases of a project from financing to management. A package deal may also involve selling or contracting out number of products and/or services at one price. See also DESIGN-BUILD PROCESS and TURNKEY PROJECT.

Panelized housing

A factory-produced package which contains parts and pieces to build a house, including wall panels, roof trusses, and a floor system. Panelized houses are of two types, those with open and those with closed wall systems. The package is sold to a builder, who then erects the house to conform to local building codes. In the U.S. building industry, a panelized house is considered a type of manufactured housing. See also MANUFACTURED HOUSING.

Parcelling-out

See LAND PARCEL.

Parkway

In the United States, a scenic limited-access highway

一揽子交易　Yīlǎnzi jiāoyì

在美国，指由一个承包商承担一项业务的全部任务的商业安排。例如，由一个合格的建造者承担一项工程从设计到施工的各阶段，包括财务和管理的全部任务。此词也可指按同一价格销售一定数量的产品或提供服务。[参见"设计兼施工"、"交钥匙工程"]

大板式住房　Dàbǎnshì zhùfáng

用工厂生产的预制构、配件，包括墙板、屋架和楼板体系等，建造的房屋。分通用和专用墙板体系两种。建房者可购买整套预制构、配件，并按当地建筑法规建造安装。在美国建筑业中，大板装配式住房是预制住房的一种。[参见"工厂预制住房"]

土地分配　Tǔdì fēnpèi

[见"地块"]

林荫公路　Línyīn gōnglù

在美国，指沿途景色宜人、限制人口的非商业性

for noncommercial traffic, usually within a park or park-like development. See also EXPRESSWAY.

Partial occupancy

See OCCUPANCY.

Pedestrian-mall

A broad strip of land, usually within a city, that is often closed to traffic and landscaped with trees, bushes, and flowers, serving as a promenade or pedestrian walk. The pedestrian mall may be part of a shopping area in a city.

Penalty and bonus clause

See BONUS AND PENALTY CLAUSE.

People's Construction Bank of China

In China, the specialized bank responsible for all financial activities related to capital construction. Its main business functions include management of capital construction budgetary expenditures and financial affairs; management of capital construction grants, loans, and account settlement; and financial supervision of development organizations and construction organizations. See also DOMESTIC LOAN.

交通用公路，一般处于公园或类似公园的地区中。[参见"高速公路"]

部分占用 Bùfèn zhànyòng

[见"占用"]

人行林荫路 Rénxíng línyīnlù

城市内禁止车行交通并以树木花草美化的较宽阔的条形地段，供散步或步行使用，也可以是城市中购物区的一部分。

惩罚和奖励条款 Chéngfá hé jiǎnglì tiáokuǎn

[见"奖罚条款"]

中国人民建设银行 Zhōngguórénmín jiànshèyínháng

中国专门管理基本建设业务的专业银行，凡与基本建设有关的一切经济活动都必须通过建设银行办理，其主要职责是管理基本建设预算支出和财务，对建设单位和施工单位进行财务监督。[参见"国内贷款"]

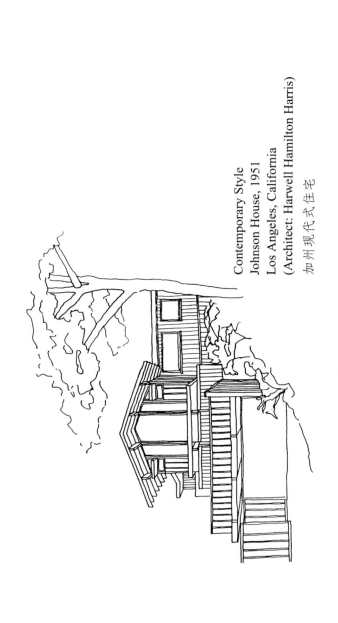

Contemporary Style
Johnson House, 1951
Los Angeles, California
(Architect: Harwell Hamilton Harris)

加州現代式住宅

HIGH-RISE BUILDINGS
三座高层建筑

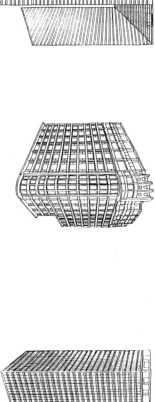

International Style
Lake Shore Drive
Apartments, 1951
Chicago, Illinois
(Architect: Mies Van der Rohe)

国际型

Chicago Commercial Style
Carson Pirie Scott Department
Store, 1899
Chicago, Illinois
(Architect: Louis Henri Sullivan)

芝加哥商业型

High-Tech Style
Pennzoil Building, 1976
Houston, Texas
(Architects: Philip Johnson and John Burgee)

高技术型

Percentage agreement

In the United States, an agreement for professional services that defines the amount of compensation as a percentage of the construction cost. See also FEE.

Percentage fee

See FEE.

Percentage of employees living with dependents

In China, the percentage of workers who reside with their dependents, as compared to the total number of workers in a factory or plant. The figure is used in calculating the amount of family housing required for employees.

Percentage rent

In the United States, rental payment for use of a commercial property consisting of a fixed amount plus a certain percentage of the receipts from the business.

Performance bond

In the United States, a bond backed by a surety, i.e., a third party, guaranteeing to a project owner that the contractor will complete construction work on time according to contract requirements. Except where prohibited by statutes, the performance

按百分比收费协议 Àn bǎifēnbǐ shōufèi xiéyì

在美国,指计取专业服务费用的一种协议,即按工程成本的一定百分比计取服务费用。[参见"费"]

百分比费用 Bǎifēnbǐ fèiyòng

[见"费"]

职工带眷比 Zhígōng dàijuànbǐ

在中国,指工厂中带家属的职工占全体职工总人数的百分比,以此测算所需职工住宅的数量。

按百分比计租金 Àn bǎifēnbǐ jì zūjīn

在美国,指使用商业性房地产所支付的租金,包括固定租金加上一定百分比的营业收入。

履行合同保单 Lǚxíng hétóng bǎodān

在美国,指由作为保证人的第三者向业主所交的保证书,保证承包人将按合同要求按时完成工程。除法令另有规定外,履行合同保单通常与人工材料费保单结合,即保证承包人将支付完成工程所需的人工材料费用。[参见"保证书"、"完工保证

bond is frequently combined with the labor and material payment bond, i.e., the guarantee that the contractor will pay for the labor and materials used for performance of the contract. See also BOND, COMPLETION BOND, and MAINTENANCE BOND.

Performance code

A code establishing design and engineering criteria, when possible without reference to specific methods of construction – as distinguished from a specification code. The standards of performance defined may be used to test the adequacy of building materials or design, or to measure pollution or nuisance levels of industrial land use activities that generate smoke, dust, fumes, or toxic wastes. Performance standards are considered to be more flexible than specification standards. See also BUILDING CODES AND STANDARDS, BUILDING MATERIAL TESTING, ENERGY PERFORMANCE STANDARD, ENVIRONMENTAL STANDARDS, PERFORMANCE STANDARD ZONING, and SPECIFICATION CODE.

Performance standard zoning

In the United States, a zoning system which defines minimum standards for nuisance and pollution

书"、"维修保证书"]

功能法规 Gōngnéng fǎguī

定出设计和工程标准的建筑法规,并不规定施工方法,以有别于规格法规,功能标准可用来鉴定建筑材料或设计的可靠程度,或测定工业烟尘及有害物质所引起环境污染的程度,功能标准较诸规格标准更为灵活。[参见"建筑法规与标准"、"建筑材料检验"、"能量效率标准"、"环境标准"、"功能标准区划"、"规格法规"]

功能标准区划 Gōngnéng biāozhǔn qūhuà

在美国,指不确定各区的用途,而仅规定控制噪声和污染最低标准的区划方法。即不是对使用而

levels rather than specific uses for each zoned area, thus regulating the impact of uses instead of the uses themselves. The system is usually applied for industrial zoning. See also ENVIRONMENTAL STANDARDS and PERFORMANCE CODE.

Permitted use

In the United States, a land use allowable in a particular district or zone as specified in the zoning ordinances for that district. See also ACCESSORY USE and ZONING.

PERT

See PROGRAM EVALUATION AND REVIEW TECHNIQUE.

Physical planning

Advance design of an area to be developed or redeveloped, taking into consideration all the existing natural and manmade physical elements of the area, including infrastructure and buildings as well as strategies for optimal development.

Piecemeal development

Construction or restoration of single buildings in or around a city without coordinated advance planning for the area. See also INFILL DEVELOPMENT and

是对使用后产生的影响作出规定。此法常用于工业区划。[参见"环境标准"、"功能法规"]

许可使用 Xǔkě shǐyòng

在美国,指在一个地区或区划范围内,按其区划法令所允许的用途使用土地。[参见"附带用途"、"区划"]

计划评审法(缩写) Jìhuà píngshěn fǎ(suōxiě)

[见"计划评审法"]

实体规划 Shítǐ guīhuà

为开发或改造一个地区而预先做出的设计,把现有一切自然和人为的物质条件纳入规划,加以全面考虑,包括基础设施、房屋建筑、最佳开发战略等。

零星建筑 Língxīng jiànzhù

在城市或其周围地区建造或修复一些事先并无统一规划的单体建筑物。[参见"填空性建设"、"见缝插针"]

"SQUEEZE-IN DEVELOPMENT."

Pilot project

In the United States, a trial program on a limited scale to test the feasibility of a particular concept, policy, or system, and to establish guidelines and parameters for development of more extensive programs of a similar type in the future. Also called a demonstration project.

"Pipeline"

In American usage, the collection of sequential stages or steps involved in a production or administrative process. The meaning of the term varies somewhat according to the process involved. For example, in government "pipeline" may refer to the "paper" process of moving an application or request through a series of reviews and clearances before it is approved, while in the commercial sector the term may be applied to actual production, i.e., processing a manuscript through to a finished book. A product or project being processed is said to be "in the pipeline."

Planned unit development (PUD)

A housing subdivision planned as a whole, including residences, roads schools, and commercial and industrial areas, and with open spaces and

试点项目 Shìdiǎn xiàngmù

在美国，指一种小规模的试验性项目，用以试验一项概念、政策或体系的可行性，并为以后在广泛规模上的发展取得指导性经验，又称"示范工程"。

流水线 Liúshuǐxiàn

在美国的用法中，指生产或行政管理过程中的一系列顺序或步骤。其含义随过程的不同而异。例如，政府"流水线"可指一项请求在被批准前所要经过的一系列公文审理的运行过程。而在商业界，则可指实际生产过程。如从整理手稿到书籍出版的过程，正在处理中的产品或工程即可说是"正在流水线上"。

有规划的地段建设 Yǒuguīhuàde dìduàn jiànshè

把划成小块住房用地的地段作为一个完整的住宅区来规划，其中包括住宅、道路、学校、商业及工业区，以及供全体居民使用的空地和娱乐设施。居住

recreational facilities for the use of all residents. Individual living units are owned privately, and responsibility for common facilities is vested in a homeowner's association. Maintenance and upkeep is financed by occupants through a legally established fee system. See also CLUSTER DEVELOPMENT, COMMON-AREA MAINTENANCE, and COMMUNITY ASSOCIATION.

Planning by object

A design or planning technique that determines the course of action best suited to achieve specific, predefined goals.

Planning commission

A local, state, or regional public agency charged with preparation and adoption of comprehensive long-term development plans for the land within the region, e.g., a city planning commission. See also CITY PLANNING COMMISSION and REGIONAL PLANNING COMMISSION.

Planning Programming Budgeting System (PPBS)

In the United States, a budgeting technique for management and decision-making in government which organizes the budget by programs of the budgeted unit (outputs) rather than by expenditures

单位归私人所有，公用设施则由房主协会负责。居民根据法定收费办法交纳管理维修费。[参见"组团式建设"、"公用部分的维护"、"社区协会"]

按目标制订规划　Àn mùbiāo zhìdìng guīhuà

为确定如何能最好地达到某一具体的预定目标而进行规划或设计的技术。

规划委员会　Guīhuà wěiyuánhuì

负责制订并采纳土地使用的综合性长期发展规划的州、地方或区域级公共机构，例如城市规划委员会。[参见"城市规划委员会"、"区域规划委员会"]

编制预算计划系统　Biānzhì yùsuàn jìhuà xìtǒng

在美国，指在政府中用于管理和决策的一种预算方法，即根据预算单位的计划（产出量）而不是根据材料、维修、人工等费用开支（投入量）来编制预算。此种方法还可根据实际所需年限做出

for material, maintenance, and personnel (inputs). The method also extends the budgetary estimates far enough into the future to demonstrate the full financial and other resource requirements for completion of the program, rather than focusing on the usual legislative budgeting period of one or two years.

Plat

In the United States, a detailed plan or map showing the proposed or existing lots or divisions, special features, and use areas of a piece of land already developed or under development.

Plot ratio

See FLOOR-AREA RATIO.

Pluralism

A condition in which different ethnic, religious, and social groups within a society maintain their separate cultural values and practices.

Police substation

In China, the grass roots unit of the national public security system, located at the subdistrict level. The police substation is responsible for such matters as public security and domicile registration. The local

预算，以保证预算单位获得财力及其他资源来完成计划，而不是采用法定预算期以一年或两年为限的预算方法。

地段图 Dìduàntú

在美国，指一种详细规划图或地图。图中标明拟议中的或现有的土地划分，其特征及其中已建或在建部分。

容积率 Róngjīlǜ

[见"建筑面积比"]

多元化社会 Duōyuánhuà shèhuì

指在一个社会中，具有不同种族、宗教和社会背景的集团各自保持其文化价值与实践的状况。

派出所 Pàichūsuǒ

中国国家公安系统最基层的单位，负责治安、户籍管理等工作。派出所受区公安分局领导，区公安分局受市公安局领导。[参见"户籍"、"户口簿"、"户口登记"]

police hierarchy includes the public security bureau at the municipal level, the public security station at the city district level, and the police substation. See also DOMICILE REGISTER, RESIDENCE CERTIFICATE, and RESIDENT REGISTRATION.

Polynucleated city

According to the multiple nuclei concept first suggested by R.D. McKenzie, an urban area with multiple centers of commercial activity, each with specialized functions.

Population density

Ratio between the number of individuals and the surface of the area on which they live. See also RESIDENTIAL DENSITY.

Population distribution

The way in which the inhabitants of a given region, area, or country are spread across the area, based on statistical estimates.

Population growth from migration

As defined in China, the expansion of the urban population as the result of factors other than natural population growth, primarily internal migration and/or job transfers, and, hence, of resident registration

多中心城市 Duōzhōngxīn chéngshì

根据 R.D. 麦肯齐首次提出的多中心概念,指一个城市地区内有多个各具特定功能的商业活动中心。

人口密度 Rénkǒu mìdù

人口数与所在地区面积之比。[参见"居住密度"]

人口分布 Rénkǒu fēnbù

指根据统计得出的某一区域、地区或国家内居民的分布状况。

人口机械增长 Rénkǒu jīxiè zēngzhǎng

在中国,指城市中人口除自然增长外,因其他原因的增长,例如因人口迁移或工作调动而将户口从一地迁至另一地所造成的人口增长。[参见"人口自然增长"]

from one place to another. See also NATURAL POPULATION GROWTH.

Postoccupancy survey

In the United States, systematic collection and analysis of data from former occupants of rental property to determine areas of satisfaction and dissatisfaction with facilities and services.

Poverty level

In the United States, the individual or family income amount officially designated as the minimum required to maintain an adequate standard of living. Definitions vary by family size and metropolitan area; figures are updated as the consumer price index changes. For example, a family of four was considered "poor" in the United States in 1983 if it had a cash income of less than $10,178.

PPBS

See PLANNING PROGRAMMING BUDGETING SYSTEM.

Prefecture

In China, the subdivision of a province or an autonomous region governed by a subunit of the provincial or regional authorities. The prefecture

使用后调查 Shǐyònghòu diàochá

在美国,指对于曾租用房地产的住户进行系统地数据搜集和分析,以确定居民对设施与服务满意与否。

贫困线 Pínkùnxiàn

在美国,指由官方规定的个人或家庭为保持最低生活水平所需的收入金额。因都市地区和家庭人口数而异,并随消费价格指数的变化而修订。例如,在美国,1983年时,10178美元以下的四口之家为"贫困家庭"。

编制预算计划系统(缩写) Biānzhì yùsuàn jìhuà xìtǒng (suōxiě)

[见"编制预算计划系统"]

地区 Dìqū

在中国,由省、自治区在其管辖范围内设立派出机构的地区,一般包括若干市、县。

usually consists of a number of municipalities and counties.

Preliminary design

In China, the first phase of architectural design. Preliminary design is based on the approved feasibility report and the design program for the project, as well as reliable basic data. This phase involves preparation of preliminary plans, specifications, and preliminary estimates for the project to be constructed. After approval, the preliminary design, with preliminary estimates, will be the basic documentation for ascertaining the amount of investment required, preparing an investment plan for fixed assets, signing the construction contract, controlling appropriation of funds, preparing for construction, and ordering essential materials and equipment, as well as developing detailed designs. See also DESIGN PHASE.

Preliminary estimate

In China, a constituent part of design documentation showing the estimated cost of a construction project calculated on the basis of basic design, norms for preliminary estimates, and norms for expenses. An approved preliminary estimate is applied in

初步设计　Chūbù shèjì

在中国，指设计工作的第一阶段。根据批准的可行性研究报告、设计任务书和可靠的设计基础资料，编制拟建工程的方案图、说明书和概算。经送审批准的初步设计和总概算，是确定建设项目的投资额、编制固定资产投资计划、签订工程合同、控制工程拨款、进行施工准备、组织主要设备订货以及进行技术设计（或施工图设计）等的依据。[参见"设计阶段"]

设计概算　Shèjì gàisuàn

在中国，指设计文件的组成部分，根据初步设计、概算定额及费用定额概略地计算出来的工程项目的概算价值。经送审批准的设计概算是确定建设项目的投资额、编制计划、签订合同及控制拨款的依据。[参见"概算定额"]

determining the investment for construction, preparing the construction program, signing the contract, and appropriating funds. See also NORM FOR PRELIMINARY ESTIMATES.

Primary mortgage market

In the United States, the sector of home finance in which first mortgages or trust deeds originate. Primary mortgage lenders, e.g., savings and loan institutions, make loans to property buyers and service the loans, which can then be held by the lender or sold to investors in the secondary mortgage market. See also MORTGAGE, MORTGAGE BANK, MUTUAL SAVINGS BANK, SAVINGS AND LOAN ASSOCIANTION, and SECONDARY MORTGAGE MARKET.

Primary population

In China, a part of the labor population, i.e., persons who are employed by industrial and transportation organizations, as well as those who are employed by administrative, financial, cultural, and educational organizations at the level above local government. The size of the primary population is a decisive factor in the size of the city. See also LABOR POPULATION and SERVICE POPULATION.

初级抵押贷款市场 Chūjí dǐyā dàikuǎn shìchǎng

在美国,指进行首次抵押或信托贷款的房产金融界。初级抵押贷款者,例如储蓄和贷款机构,为购买房地产筹措并支付贷款,贷款者可持有抵押债权或在二级抵押市场中转卖给投资者。[参见"抵押贷款"、"抵押贷款银行"、"互助储蓄银行"、"储蓄贷款协会"、"二级抵押贷款市场"]

基本人口 Jīběn rénkǒu

在中国,指劳动人口中的一部分,即在工业、交通运输单位及不属于地方性的行政、财经文教等单位中工作的人员。这部分人口对于城市规模起决定性作用。[参见"劳动人口"、"服务人口"]

Primary wastewater treatment

The first stage of purifying wastewater, involving removal of discrete, suspended, settleable, and floating solids from raw sewage by processes for screening, grinding, and settling. See also ADVANCED WASTEWATER TREATMENT, SECONDARY WASTEWATER TREATMENT, SEWAGE DISPOSAL SYSTEM, WASTE MANAGEMENT, and WASTE UTULIZATION.

Primate city

A single large city in a society with a low level of urbanization.

Prime contract

In the United States, a contract between the owner and contractor for provision of products or services, or for construction of a project or a portion thereof. The prime contractor may then subcontract out portions of the work. In common usage, the meanings of prime contract and general contract overlap. See also CONTRACT, CONTRACTOR, GENERAL CONTRACT, and SUBCONTRACT.

Prime interest rate

In the United States, the interest rate charged by financial institutions for loans to their best customers,

污水一级处理　Wūshuǐ yījí chǔlǐ

指净化污水的第一阶段,包括通过过滤、磨碎和沉淀除去污水中的浮游物和悬浮体等的全过程。[参见"污水深度处理"、"污水二级处理"、"污水处理系统"、"废物管理"、"废物利用"]

初级城市　Chūjí chéngshì

都市化水平低的单一的大城市。

直接承包合同　Zhíjiē chéngbāo hétóng

在美国,指业主与承包人之间签订的有关提供产品、服务、承包全部或部分工程项目的合同。直接承包人又可将部分工作量分包出去。在一般用语中,直接承包合同与总承包合同的含义相近。[参见"合同"、"承包人"、"总承包合同"、"分包合同"]

优惠利率　Yōuhuì lìlǜ

在美国,指金融机构向最佳主顾,即信用最高的实业界借户,提供贷款时所收的利息率。

i.e., business borrowers with the highest credit ratings.

Principal use

In the United States, the primary use of any land parcel, usually regulated by zoning ordinances.

Priority lane

In the United States, a highway lane reserved for the exclusive use of buses and car pool vehicles to encourage citizen use of those means of transportation and to increase effective highway capacity. See also CAR POOL.

Privacy

State of being away from others, alone and undisturbed. In residential areas, privacy includes both a desirable degree of seclusion, i.e., visual privacy, and peace and quiet, which means requiring protection from noise.

Private lot

1. In China, land occupied by a private house, mainly in rural areas. The individual who resides there has tenure but does not own the land.
2. In the United States, land owned by a private individual.

主要用途 Zhǔyào yòngtú

在美国,指由区划法令规定的某一地块的基本用途。

优先车道 Yōuxiān chēdào

在美国,指公共汽车或多人合乘车辆专用的公路车道,用以鼓励市民使用此种交通工具,提高公路的有效运输量。[参见"合乘"]

私蔽性 Sībìxìng

离开他人独处而不受干扰。在居住区中,既包括视觉上的不受干扰,即合适的隐蔽;也包括平静安宁、不受噪声的干扰。

宅基地 Zháijīdì

1. 在中国,主要指农村中私有住宅所占的土地。宅主对土地仅有使用权,并无所有权。

2. 在美国,指私人所拥有的土地。

Private sector

In open market economics, the portion of the economy dominated by privately owned business and industry. See also PUBLIC SECTOR.

Privately aided public housing

In China, an approach to housing finance which seeks to raise funds from private citizens to alleviate the housing shortage. In housing developments of local governments, institutions, or enterprises, individuals in need of housing can make contributions in the form of advance rental payments in return for priority in housing allotment. Housing thus built is publicly owned. See also PUBLICLY AIDED PRIVATE HOUSING.

Production management

See MANAGEMENT STRUCTURE.

Productivity

In economics, the output of process measured in relation to the labor expended. For an industry or economy, productivity serves as a measure of overall efficiency, comparing system inputs (materials, labor, and overhead) to outputs (products and/or services).

Program Evaluation and Review Technique (PERT)

A mathematical method for advance determination

私有部分 Sīyǒu bùfèn

在开放型市场经济中,指由私人工、商业控制的那部分经济。[参见"公有部分"]

公建民助 Gōngjiàn mínzhù

中国城市中向群众集资建房以缓和住房短缺的一种措施。地方政府或企、事业单位在建设住房时,职工可用抵租形式投入部分资金,以取得优先居住权。住房建成后,产权归地方政府或企、事业单位所有。[参见"自建公助"]

生产管理 Shēngchǎn guǎnlǐ

[见"管理结构"]

生产率 Shēngchǎnlǜ

在经济学中,指生产过程中所花的劳动力与所取得的产量之比。工业或经济部门通过系统地比较输入(材料、劳动力和管理费)与输出(产品和/或服务)计算出生产率,作为衡量全面效率的尺度。

计划评审法 Jìhuà píngshěn fǎ

为能提前确定完成一项工程所需的最少时间而采

of the minimum time needed to complete a proposed project. The most important element in PERT is the activity-event network plan which shows graphically the component tasks, their interrelationships, and the sequence they must follow. See also CRITICAL PATH METHOD.

Programming phase

A predesign phase in which the architect or a design organization assists the project owner in defining the basic design concept and requirements for the project, and in writing a program. See also DESIGN PROGRAM.

Progress payment

Payment under a contractual agreement on the basis of work progress, either for work actually completed or at a defined percentage rate. In China, payment is for work actually completed and is made though the People's Construction Bank of China by the owner of a construction project to the construction organization.

Project budget

The sum established by the owner for the entire project, including the construction budget, land costs, equipment costs, financing costs, compensation for

用的一种数学方法,其核心是活动—事件网络图,用以显示工作的构成、顺序及其相互关系。[参见"关键线路法"]

计划阶段 Jìhuà jiēduàn

设计工作开始前的一阶段。在此期间由建筑师或设计单位协助业主确定基本设计构想和对工程的要求,并写成计划。[参见"设计任务书"]

工程进度款 Gōngchéng jìndùkuǎn

按照合同规定根据工程进度所付的款项。可按实际完成的工程量付款,亦可按规定的比例付款。在中国,是根据实际完成的工作量,由建设单位通过中国人民建设银行拨付给施工单位。

建设项目投资预算 Jiànshè xiàngmù tóuzī yùsuàn

业主所确定的建设项目的总投资额,其内容包括工程投资预算、土地、设备、财务、专业服务及其他确定或估计的费用。[参见"工程预算"、"不

professional services, contingency allowances, and other similar established or estimated costs. See also CONSTRUCTION BUDGET and CONTINGENCY ALLOWANCE.

Project in preparation

One of the main statistical indices of capital construction in China. This index refers to a construction project which has already been approved and listed in the preparation stage prior to commencement of regular construction work.

Project manager

In the United States, the individual assigned the responsibility for carrying out and managing all or specified portions of a project, whether construction or some other type of service. See also CONSTRUCTION MANAGER.

Project owner

See OWNER.

Project/item under construction

A major statistical index of capital construction in China, referring to a construction project or portion of a project presently being built.

可预见费"]

筹建项目 Chóujiàn xiàngmù

在中国,是基本建设主要统计指标之一,指已被批准列入国民经济与社会发展计划而尚处于工程开工前施工准备阶段的建设项目。

工程主任 Gōngchéng zhǔrèn

在美国,被指定对全部或部分工程或其他服务项目的实施负责的人。[参见"施工经理"]

建设项目业主 Jiànshè xiàngmù yèzhǔ

[见"业主"]

施工项目 Shīgōng xiàngmù

在中国,是基本建设主要统计指标之一,指处于施工阶段的建设项目或工程项目。

Property insurance

In the United States, protective coverage of real or personal property by contract against loss or damage incurred directly, as a result of liability, or through dishonesty or failure of others to perform their duties. Coverage of mortgaged real property is usually a requirement of the lender. See also COINSUREANCE.

Province

In China, the highest level local administrative subdivision, with its own government. The provinces report to the central government.

Provincially administered municipality

See MUNICIPALITY.

Public hearing

In the United States, a preannounced meeting that is open to the public and allows citizens to voice their views on public undertakings of general concern. Public hearings are often required before adoption or implementation of a master plan.

Public housing

In the United States, dwellings for low-income families, built and operated by local housing

财产保险 Cáichǎn bǎoxiǎn

在美国,指用合同形式对不动产和动产由于债务、他人失信或不履行义务所直接造成的损失、损坏所做的合同保险。贷款者通常都要求受抵押的财产具有保险。[参见"共同保险"]

省 Shěng

在中国,中央政府下属的地方最高一级行政区域单位,设省政府。

省辖市 Shěngxiáshì

[见"市"]

公众意见听取会 Gōngzhòng yìjiàn tīngqǔhuì

在美国,指在执行某项公共计划前预先宣布的、对公众开放的听取意见会议,以便让市民对普遍关心的公共事宜发表意见。在采纳或实施一个总体规划前,一般需召开此种会议。

公共住房 Gōnggòng zhùfáng

在美国,指由联邦政府提供财政支持,由地方住房主管部门建造和管理的低收入家庭住房。在全

authorities with financial backing from the federal government. Public housing represents less than 2 percent of the existing housing stock. See also ASSISTED HOUSING and HOUSING MANAGEMENT.

Public improvements

Development, upgrading, or maintenance of a community's facilities or infrastructure, e.g., streets, utilities, and schools, designed to meet the needs of citizens.

Public institution

An organization that draws all or part of its funds from a government and is nonprofit, e.g., a school, a hospital, or a research institute. In China, the public institution is under the direction of the government and is referred to simply as an institution; in the United States, it may or may not be government-administered, depending on its charter and level of government funding.

Public participation

See CITIZEN PARTICIPATION.

Public relations

Activities undertaken by an individual or organization

部住房储存量中,公共住房所占比例不足2%。
[参见"公助住房"、"住房管理"]

公共设施的改进 Gōnggòngshèshī de gǎijìn
对社区设施或基础设施,例如街道、市政公用设施和学校等,进行开发、提高或维修,以满足市民的需要。

公共机构 Gōnggòng jīgòu
从政府获得全部或部分资金的非营利性组织,例如学校、医院、研究机构等。在中国,这类机构一般称为"事业单位",受政府领导;在美国,公共机构是否受政府领导,取决于它的执照的性质和政府资助的程度。

公众参与 Gōngzhòng cānyù
[见"市民参与"]

公共关系 Gōnggòng guānxì
个人或组织为提高信誉和树立正面形象而采取的

to promote general good will and a positive public image. Public relations may include distribution of information and promotional materials, public lectures, or any other form of activity that provides publicity for the individual or organization. "PR work," as such activities are called in everyday speech, is closely related to the advertising business.

Public sector

In free market economics, the part of the economic structure responsible for government activities and financed primarily by tax levies or other funds derived from the public. See also PRIVATE SECTOR.

Public utilities

A general term for urban facilities such as the water supply, sewage system, light and power systems, communications, heating systems, and gas supply. In the United States, the utility providing the service is frequently privately owned but is closely regulated by the government, especially with regard to rates. Government controls for this industry are particularly stringent. See also INFRASTRUCTURE.

Public works

In the United States, fixed facilities or installations

各种行动,包括散发有关个人或组织发展前景的资料、发表公开演讲或其他各种形式的有利于提高个人或组织的知名度的行动,这些在日常用语中被称为"公共关系工作"的活动与广告事业有密切关系。

公有部分 Gōngyǒu bùfèn

在自由市场经济中,指负责政府的活动,以税收和其他公益收入为主要经济来源的那部分经济成分。[参见"私有部分"]

市政公用设施 Shìzhèng gōngyòng shèshī

城市供水、排水、照明、动力、通信、供热、燃气等设施的统称。在美国,这种设施虽然多为私营,但由政府严加管理,尤其在收费方面,政府对这一行业的控制特别严格。[参见"基础设施"]

公共设施 Gōnggòng shèshī

在美国,指由政府所有,以公款建造供公众享用

such as roads or playgrounds constructed and owned by the government with public funds for public use or enjoyment. See also INFRASTRUCTURE.

Public-private partnerships

In the United States, cooperative efforts of government with the private sector, especially business and industry, in specific area of development or common concern such as national defense, improvement of health care and education, and elimination of poverty. Public-private partnerships are now a key factor in economic development strategies linked to urban revitalization.

Publicly aided private housing

In China, an approach to housing finance designed to alleviate the urban housing shortage through combined use of public and private funds. Under this system, individuals who need to build new homes or to make additions or improvements to their existing houses raise the major portion of the funds themselves, and the organizations for which they work make up the difference. The funds are repaid in an agreed period of time. The house thus built is privately owned. See also PRIVATELY AIDED PUBLIC HOUSING.

的固定设施或装备,如道路、游戏场等。[参见"基础设施"]

公私合作 Gōngsī hézuò

在美国,指政府和私有部分尤其是工商企业,在某些建设领域或共同关心的方面,如国防、改善医疗卫生和教育以及消除贫穷等进行合作。公私合作已成为与城市复苏相关联的经济发展战略上的关键因素。

自建公助 Zìjiàn gōngzhù

在中国,指调动公家和私人的资金建房以缓和城市中住房短缺的一种措施。私人新建住房或对其原有住房加建或翻建时,由本人出大部分资金,不足部分由其工作单位资助,建房人按期偿还。建成后的房屋产权归私人所有。[参见"公建民助"]

PUD

See PLANNED UNIT DEVELOPMENT.

Punch list

In the United States, a list of discrepancies in building plans or of construction flaws recorded by the original architect during final inspection of the structure and requiring correction for completion of the work as specified in the contract. See also FINAL ACCEPTANCE and NONCONFORMING WORK.

有规划的地段建设(缩写) Yǒuguīhuàde dìduàn jiànshè (suōxiě)

[见"有规划的地段建设"]

建设缺陷清单 Jiànshè quēxiàn qīngdān

在美国,指记录与原设计不符的各种问题或施工缺陷的清单。由负责该项工程的建筑师在对建筑物作最终检验时记录在案,并按合同规定要求承包单位予以改正。[参见"最终验收"、"不符合同工程"]

Q

Quadruplex house

In the United States, four living units constructed with common walls and generally with a private entrance for each unit. This combination of units provides a reasonably priced alternative to the single-family detached house. See also ATTACHED HOUSE.

Quality control

The ongoing process of ensuring an acceptable standard of quality in services or goods by critically examining the initial system or product design and monitoring delivery or production at various stages. The responsibility for quality control starts with the manufacturer or producer, and extends frequently to a testing or inspection agency and, in special cases, to a third agency that validates testing or inspection. Adjustments are made in the design or procedures as needed. See also BUILDING MATERIAL TESTING, CERTIFICATION OF MATERIALS AND PRODUCTS, MONITORING, and TECHNOLOGY ASSESSMENT.

Quantity survey

1. In the United States, a list of all types and quantities of materials and equipment required for a construction project. Also called a bill of quantities.

四联式住宅 Sìliánshì zhùzhái

在美国，指有共用界墙且每单元有独用入口的四个居住单元。这种组合单元住房的价格较独立式住宅便宜。[参见"毗连住宅"]

质量控制 Zhìliàng kòngzhì

为保证产品或服务质量达到合格标准而对初次采用的体系或产品设计进行检查，并在各个阶段对生产或运输进行监督。质量控制最初是生产者的责任，而后扩大到测试单位，在特殊情况下还可由第三方作独立的检验。根据检验需要可对设计或生产步骤做出调整。[参见"建筑材料检验"、"材料与制品鉴定"、"施工监督"、"技术评定"]

1. 工程用料与设备清单 Gōngchéng yòngliào yǔ shèbèi qīngdān

在美国，指一个施工项目所需的全部材料和设备

2. In the United States, a detailed listing and mapping of vacant land available for use or of available dwelling units. See also LAND USE SURVEY.

的类型与数量清单。

2. **空地或空房调查表** Kòngdì huò kōngfáng diàochábiǎo 在美国，指可供使用的空地或可供居住的空房的调查清单和方位图。[参见"土地利用调查"]

R

Rapid transit

See MASS TRANSIT.

Rate of return

The amount of money earned on an investment, usually expressed as a percentage of the amount of funds invested.

Real estate

1. In the United States, land with its permanent structures, infrastructure, natural resources such as water and minerals, and any rights or interests in rights associated with ownership of the land. Also called real property in a legal context.
2. In the United States, the area of business concerned with land and property sales and purchase transactions.

Real property

See REAL ESTATE.

Reclamation

See LAND RECLAMATION.

Reconstruction

Action of constructing one or several new buildings to replace buildings which have suffered damage.

快速运输 Kuàisù yùnshū

[见"大量客运"]

利润率 Lìrùnlǜ

指用于投资的资金所赚取的利润,通常以投资金额的百分比表示。

房地产 Fángdìchǎn

1. 在美国,指土地及其上部的永久性建筑物以及基础设施和诸如水和矿藏等自然资源,还包括与土地所有权所有关的任何权利或利益。在法律上称为不动产。

2. 在美国,又称经营房地产买卖的商业界。

不动产 Bùdòngchǎn

[见"房地产"]

整治 Zhěngzhì

[见"土地整治"]

重建 Chóngjiàn

建造新建筑代替已损坏的原有建筑物,亦可指对因受灾而损坏的建筑物进行的修复。

Reconstruction may also involve repair of damage to buildings caused by a disaster.

Record drawings

Construction drawings revised to show significant changes made during the construction process, usually based on mark-up prints, drawings, and other data furnished by the contractor or the architect.

Recycling

1. Recovery and reuse of basic materials from such products as newspapers, aluminum cans, and bottles, as well as building materials such as bricks. See also RESOURCE RECOVERY.
2. In the United States, conversion of abandoned, vacant, or obsolete buildings that are structurally sound so that they can be productively reused.

Redevelopment

Improvement of cleared or undeveloped land, including erection of buildings and other facilities, by developers, usually in an urban renewal area. See also URBAN REDEVELOPMENT and "WRITE-DOWN COSTS."

Redlining

In the United States, a practice of financial

竣工图 Jùngōngtú

标明在施工过程中做了重要变更的施工图,常以承包人或建筑师所提供的带标记的图纸或其他资料为依据。

再利用 Zàilìyòng

1. 从废旧产品中,如报纸、铝罐头、玻璃瓶及砖制品等建筑材料,回收基本材料并加以利用。[参见"资源回收"]

2. 在美国,指对结构尚坚固但已被废弃、关闭或陈旧的房屋加以修复改造,重新使用。

改造 Gǎizào

通常指在城市更新地区内,对已清理或尚未开发的土地加以改善,包括由开发者建造房屋和其他设施。[参见"城市改造"、"减记成本"]

红线注销 Hóngxiàn zhùxiāo

美国金融机构的一种做法,即指出城市中某些地

institutions whereby certain area of a city are deemed ineligible to receive loans because of the high risk of default. The result is further depression of business activity in already distressed districts. See also UNDERWRITING.

Refinance

In the United States, restructuring of a debt (personal, corporate, or national) by taking out a new loan to replace an old loan. The terms of a new loan (e.g., interest rate, amortization schedule, etc.) are more advantageous to the debtor than the terms of the previous loan. In real estate, one can speak of "refinancing a mortgage" on the same property. See also LOAN TERMS and MORTGAGE.

Regional planning commission

In the United States, an agency established by states to undertake regional planning, typically governed by a board consisting of government-appointed experts, local governmental representatives, or interested citizens. The board approves regional plans and other studies prepared by professional staff, reviews applications for federal grants by local governments, and provides advisory services to the business community and to local governments. The power of commissions varies widely, with many of them

区由于有高度的违约风险而不宜接受贷款，其结果是使已经不景气地区的商业活动进一步萧条。[参见"风险担保"]

重筹资金 Chóngchóu zījīn

在美国，指以新贷款代替老贷款重组个人、公司或国家的债务。新贷款的条件（例如利率、分期偿还的进度等）与原贷款相比，对债务人更为有利。在房地产界，指以同一产业重筹抵押贷款。[参见"借款条件"、"抵押贷款"]

区域规划委员会 Qūyù guīhuà wěiyuánhuì

在美国，指由州政府设立的负责区域规划工作的机构，常由包括政府指派的专家、当地政府代表及关心规划工作的公民组成的委员会领导。委员会有权批准区域规划及由专业人员提出的其他研究报告，审查当地政府向联邦政府提出的拨款申请，并向当地政府或商业界提供咨询。这种机构的权力因地而异。大多数委员会仅对下级行政机构如县或城市有咨询权。[参见"规划委员会"]

having only advisory power over lower jurisdictions, such as counties or cities. See also PLANNING COMMISSION.

Registered resident

In China, a resident and member of a household approved and registered by the local police substation or township government. Registered residents may be classified as permanent or temporary, urban or rural, or collective (single individuals living in dormitories of organizations). See also DOMICILE REGISTER, RESIDENCE CERTIFICATE, RESIDENT REGISTRATION, TRANSFER OF DOMICILE REGISTRATION, and URBAN INHABITANT.

Regulations

See BUILDING CODES AND STANDARDS.

Rehabilitation

Restoration to good condition of deteriorating buildings, neighborhoods, and public facilities in an urban area. See also ADAPTIVE RE-USE and URBAN REDEVELOPMENT.

Relocation

1. Movement of residents from blighted urban areas to other locations prior to major redevelopment.

户口 Hùkǒu

在中国,经当地派出所或乡政府核定登记的住户和人口的总称。户口分常住户口和临时户口、城市户口和农村户口;此外,居住在单位宿舍中的单身职工,共立一户,称为集体户口。[参见"户籍"、"户口簿"、"户口登记"、"转户"、"城市居民"]

规程、规定 Guīchéng、guīdìng

[见"建筑法规与标准"]

复兴 Fùxīng

在城市地区,将日趋衰败的建筑物、邻里或公共设施修复到良好状况。[参见"适应性再使用"、"城市改造"]

搬迁 Bānqiān

1. 在城市大规模改造工作开始前,将居民由衰落区迁移到其他地区。

2. Transfer of operations or employees of an organization from an old location to a new or different one. See also DISPLACEMENT OF FAMILIES, JOB DISPLACEMENT, and RELOCATION HOUSEHOLD.

Relocation household

In China, a household that is rehoused in another dwelling when its original dwelling is demolished for new development. See also COMPENSATORY REPLACEMENT OF DEMOLISHED HOUSING, RELOCATION, and RELOCATION HOUSING.

Relocation housing

In China, housing provided as temporary accommodations for households relocated when their dwellings are demolished. See also RELOCATION HOUSEHOLD.

Remodeling

See HOME IMPROVEMENT.

Remote sensing

Collection of information about the physical structure of the earth's surface, using such long-range techniques as aerial photography and various types of measurement from satellites. Remote sensing has

2. 亦可指某一单位的业务活动或雇员从原来地点迁到新的地点。[参见"家庭迁移"、"就业转移"、"拆迁户"]

拆迁户 Chāiqiānhù

在中国，因进行新的建设而需将原住房拆除并另行分配住房的住户。[参见"原拆原建"、"搬迁"、"周转房"]

周转房 Zhōuzhuǎnfáng

在中国，指为原住房需拆除的住户所提供的临时住房。[参见"拆迁户"]

翻修 Fānxiū

[见"住房改善"]

遥感技术 Yáogǎn jìshù

利用诸如高空照像和多种卫星测量等远距离技术收集地球表面物理结构的资料。此项技术是制订土地利用规划的重要工具。[参见"航空测量"]

proved to be a valuable tool for land use planning. See also AERIAL SURVEY.

Renovation

See RESTROATION.

Rent control

In the United States, limits placed by a local governmental authority on the permissible increase in the rent amount for dwelling units. Only a small percentage of U.S. cities and towns have rent control ordinances.

Rental housing

A house or apartment for which a tenant pays a landlord or owner agreed-to rental charges, usually on a monthly basis, for the right of occupancy. In the United States, a formal lease is normally signed defining rental terms for a fixed period. Utilities may or may not be part of the rent. See also LEASE and MARKET RENT.

Replacement cost

The current cost of replacing a fixed asset that has been lost or damaged with one of equal quality and effectiveness.

更新 Gēngxīn

[见"修复"]

房租控制 Fángzū kòngzhì

在美国,指地方政府主管部门对住房租金增长幅度所作的限制。在美国,仅有少数市镇有租金控制法令。

租用住房 Zūyòng zhùfáng

指住户向房产主租用住宅或公寓。住户每月按双方商定的金额支付房租以取得居住权。在美国,在正式租约中通常写明一定租期内的租金条件。设备是否包括在租金内则根据情况各有不同。[参见"租约"、"市场租价"]

重置成本 Chóngzhì chéngběn

重置一项与已遗失或损坏的固定资产具有相同质量和功能的固定资产的现时成本。

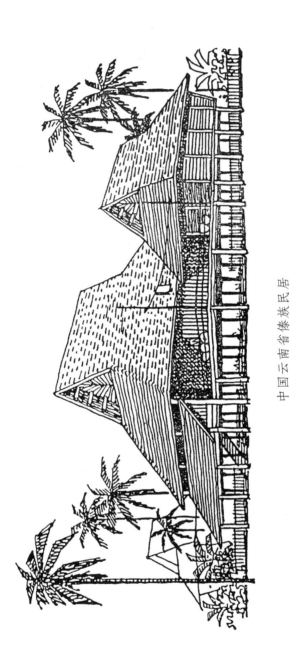

中国云南省傣族民居

Vernacular dwelling of the Tai Nationality in Yunnan Province

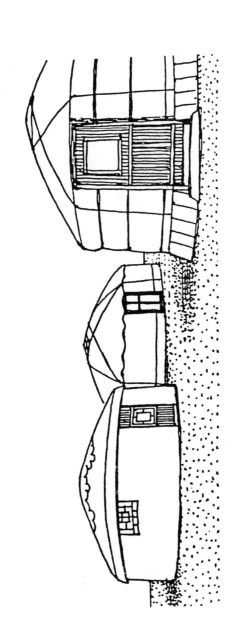

中国内蒙古自治区毡包式民居
Tent in Inner Mongolia Autonomous Region

Replacement housing

Housing units constructed to replace dwellings demolished in the course of urban renewal.

Request for Proposal (RFP)

A document issued by a U.S. government agency enumerating in detail the requirement for a proposed project or study and soliciting bids from interested private-sector contractors for performing the work required and supplying the necessary equipment or studies. See also BIDDING DOCUMENTS.

Residence certificate

In China, a certificate that is issued by the police substation or township government to a household, attesting that each member of the household is officially registered in the neighborhood and recording the basic personal information from the domicile register maintained at the substation or by the township government. See also DOMICILE REGISTER, HOUSEHOLD, POLICE SUBSTATION, REGISTERED RESIDENT, RESIDENT REGISTRATION, and URBAN INHABITANT.

Resident architect

In the United States, an architect at a job site who supervises the work and protects the owner's interests

拆迁户住房 Chāiqiānhù zhùfáng

在城市更新过程中,为原住房拆除的住户另建的住房。

招标通告 Zhāobiāo tōnggào

由美国政府机构发出的招标文件,详细列举拟议中的工程或研究项目的各项要求,邀请有意于此的私营承包商投标承担该项工程并提供所需设备或研究工作。[参见"招标文件"]

户口簿 Hùkǒubù

在中国,由派出所或乡政府按户登记核实,发给居民的户口凭证,其中记述户籍中的主要项目。[参见"户籍"、"户"、"派出所"、"户口"、"户口登记"、"城市居民"]

驻场建筑师 Zhùchǎng jiànzhùshī

施工期间为了业主的利益驻在工地监督工程的建筑师,受顾问建筑师的领导,亦称现场建筑师。

during construction under the direction of the consulting architect. Also called a site architect.

Resident engineer

In the United States, an engineering specialist employed by the owner to represent the owner's interests on the construction site during construction work. Also called a site engineer.

Resident registration

In China, all citizens are registered either at police substations as urban residents or by township governments as rural residents, with one household as a unit. After being registered, residents are given residence certificates. See also DOMICILE REGISTER, HOUSEHOLD, POLICE SUBSTATION, REGISTERED RESIDENT, RESIDENCE CERTIFICATE, TRANSFER OF DOMICILE REGISTRATION, and URBAN INHABITANT.

Residential building

A building for the purpose of living, e.g., a house, an apartment building, or a dormitory. In China, the residential building is one of the principal subclassifications of the civil building. See also CIVIL BUILDING.

驻场工程师 Zhùchǎng gōngchéngshī

在美国,受业主聘用,代表业主利益,施工期间驻在工地监督工程的工程专家,亦称现场工程师。

户口登记 Hùkǒu dēngjì

中国公民均需以户为单位进行户口登记,城市居民在当地派出所登记,农村居民在乡政府登记,登记后发给居民户口簿。[参见"户籍"、"户"、"派出所"、"户口"、"户口簿"、"转户"、"城市居民"]

居住建筑 Jūzhù jiànzhù

供居住用的建筑物的统称,例如住房、公寓楼、宿舍等。在中国,居住建筑是民用建筑的一种。[参见"民用建筑"]

Residential density

A measure of residents or dwelling in a residential area. Gross residential density shows the number of people or dwellings per unit area over the whole residential area, including the area covered by housing sites, local public facilities, and the network of distributor roads and pedestrian paths. Net residential density indicates the number of people or dwellings for the actual area covered by housing sites only. In China, residential density is registered mainly as three indices: 1) density of registered inhabitants, 2) density of residential floor area, and 3) density of living floor area. See also DENSITY OF LIVING FLOOR AREA, DENSITY OF REGISTERED INHABITANTS, DENSITY OF RESIDENTIAL FLOOR AREA, and POPULATION DENSITY.

Residential district

1. An urban area in which dwellings and public facilities are concentrated. See also DISTRICT.
2. In China, a defined unit of urban structure which has a population of 30,000 to 50,000 in a large city and 10,000 to 20,000 in a medium-sized city, and which is composed of housing estates and housing groups at the next levels. See also HOUSING ESTATE and HOUSING GROUP.

居住密度　Jūzhù mìdù

居住区内的人口数或住房套数与居住区面积之比，分毛密度与净密度两种。毛密度是按住房用地及公共服务设施、道路和人行道用地计算；净密度则仅按住房用地计算。在中国，居住密度有三个主要指标：①居住人口密度；②居住建筑面积密度；③居住面积密度。［参见"居住面积密度"、"居住人口密度"、"居住建筑面积密度"、"人口密度"］

居住区　Jūzhùqū

1. 城市中，住房及公共服务设施集中的地区的总称。［参见"区"］
2. 在中国，特指居住区规划结构中的一个层次，由若干居住小区和住宅组团组成，其规模在大城市中为3—5万居民，在中等城市为1—2万居民。［参见"居住小区"、"住宅组团"］

Residential floor area

A general designation in China for the total floor area of all floors between the outer faces of containing walls in a residential building, including living floor area, service floor area in the dwelling units, common floor area, and the area occupied by walls. See also DENSITY OF RESIDENTIAL FLOOR AREA, DWELLING SIZE, LIVING FLOOR AREA, SERVICE FLOOR AREA, and USABLE FLOOR AREA.

Resource management

1. Control and monitoring of the use of natural resources such as timber, wildlife, coal, oil, and natural gas to preserve the environment and to prevent waste of basic materials and energy sources. See also CONSERVATION and TERRITORIAL PLANNING.
2. Supervision of the allocation and use of personnel, equipment, and building space in a private or public organization.

Resource recovery

The practice of recycling reusable materials rather than discarding them entirely, e.g., newspapers and aluminum cans. See also RECYCLING.

居住建筑面积　Jūzhù jiànzhù miànjī

在中国，指居住建筑的全部建筑面积，即各层外墙皮内的全部楼面面积，包括各居住单元内的居住面积、辅助面积、公共面积和结构所占的面积。[参见"居住建筑面积密度"、"户型"、"居住面积"、"辅助面积"、"使用面积"]

资源管理　Zīyuán guǎnlǐ

1. 对天然资源如木材、野生动植物、煤、石油和天然气等的利用所做的控制和监测，以保护环境、防止基本原料及能源的浪费。[参见"保护"、"国土规划"]

2. 在私营或公共机构中，对人员、设备与建筑空间的分配和使用所做的监督。

资源回收　Zīyuán huíshōu

对可再次使用的物质，如旧报纸、铝罐头等进行回收再用，而不是将其丢弃。[参见"再利用"]

Restoration

The process of returning a building or district to its original design, as in the case of historic districts or damaged structures. Renovation also involves rebuilding a structure but does not necessarily consider its original design. See also HISTORIC DISTRICT and HISTORIC PRESERVATION.

Retrofitting

Modifications to the structure and systems of buildings to improve their usefulness and efficiency.

Revenue financing

See BOND.

Revolutionary site

In China, a commemorative site that achieved significance in revolutionary history, with buildings and relics that warrant preservation, protection, and restoration.

Revolving fund

In the United States, a monetary account that is continually drawn on and replenished, so that a reasonable balance is maintained at all times. This fund is usually renewed by income from the activities that it finances.

修复 Xiūfù

将一幢建筑物或一个地区,如一个历史性地段或一个被破坏的结构恢复到原设计的面貌。更新则指重建一个建筑物而不一定考虑其原有设计。[参见"历史性市区"、"历史性建筑保护"]

翻新 Fānxīn

对建筑物的结构及设备系统进行改造以提高其适用性和有效性。

收入财政 Shōurù cáizhèng

[见"债券"]

革命纪念地 Gémìng jìniàndì

在中国,指在革命历史上具有纪念意义而必须予以保存、保护和修复的场所、建筑与文物等。

周转金 Zhōuzhuǎnjīn

在美国,指一笔开立户头的资金,可随时提取或存入,始终保持合理的平衡,并从其投资活动的收益中不断得到补充。

Rezoning

In the United States, amendment of existing zoning ordinances through procedures defined in state enabling acts. Rezoning permits changes in land use. See also ZONING and ZONING AMENDMENT.

RFP

See REQUEST FOR PROPOSAL.

Ribbon development

See STRIP DEVELOPMENT.

Right of occupancy

In China, the right of a cadre or worker of an organization, or that of a resident in a city, to reside in a specific dwelling allotted by the organization or the housing management office, respectively, usually evidenced by a dwelling certificate attesting that the individual is the legal user of the dwelling unit. The owner of a private house has a natural right of occupancy. See also ENTITLEMENT TO HOUSING and OBLIGATION OF ORGANIZATIONS TO PROVIDE HOUSING.

Right-of-way

1. In the United States, the strip of land on which public highways are built and railroad beds are laid.

重新区划 Chóngxīn qūhuà

在美国,指通过州的有关法案对现有区划条例作出修正,允许对土地使用作出更改。[参见"区划"、"区划修正"]

招标通告(缩写) Zhāobiāo tōnggào(suōxiě)

[见"招标通告"]

带形商业区 Dàixíng shāngyèqū

[见"条形商业区"]

居住权 Jūzhùquán

在中国,指职工或市民经企业、事业单位或城市房屋管理部门分配住房后,即具有居住权,一般以住房证为凭,以证明居住者为该房的合法使用者。私有住房的房主对其住房有当然的居住权。[参见"住房申请权"、"住房分配"]

1. 筑路用地 Zhùlù yòngdì

在美国,指铺设公路、铁路用的地。

2. In the United States, the path on which individuals may legally cross other persons' land in order to gain access to their own land. See also ACCESS RIGHT and EASEMENT.

Risk analysis

An assessment of the possible financial losses or other negative consequences that might ensue from following a particular course or courses of action. The process is generally associated with major financial or other management decisions where failure would have serious repercussions for the agency responsible for the final decision, e.g., a bank financing a multimillion-dollar urban development project or a government body issuing new seismic construction standards. Depending on the particular circumstances, any of several different methodologies can be adapted to assess the various economic, social, environmental, and other interacting factors that might jeopardize the future success of an undertaking. See also EXTERNALITIES and FORECASTING.

Rowhouse

In the United States, a narrow, usually two-story housing unit connected to a continuous building. Because of their space efficiency, rowhouses have long been popular for residential housing in

2. 通行权　Tōngxíngquán

在美国，指法律允许通过他人土地以出入自己土地的权利。[参见"出入权"、"地役权"]

风险分析　Fēngxiǎn fēnxī

指对从事一项或一系列活动可能导致的经济损失或不利后果所进行的分析。在做出重大经济或行政决策前一般都需进行此种分析，因为决策的错误可能造成严重的经济后果。例如，银行对价值数百万美元的城市开发项目提供资金或政府机构颁发新的建筑抗震标准等。分析中，可按不同情况采取不同的方法学，以评价不同的经济、社会、环境以及与事业的未来成败有关的其他因素。[参见"外在副作用"、"预测"]

联立式住宅　Liánlìshì zhùzhái

在美国，指一排由面宽较窄的两层居住单元以共用界墙联结组合的住宅。由于它的空间利用率高，长期以来是美国东部大城市中流行的住宅形式。近年来所建的联立式住宅，常称"市镇联立式住

large cities of the eastern United States. Recently constructed rowhouses are usually called townhouses and are often found in the suburbs. Traditionally, this type of housing was confined to the densely populated central city areas. See also ATTACHED HOUSE and TOWNHOUSE.

宅"。按原来传统，此种住宅多建于人口密集的市中心地区，但现在也常建于郊区。［参见"毗连住宅"、"市镇联立式住宅"］

S

Sale-leaseback

In the United States, a transaction in which used equipment or real property is purchased from one party by another and then leased by the new owner to the previous owner for an extended period. The two parties enter into this arrangement because it is financially rewarding for both, particularly with regard to tax benefits. See also TAX SHELTER.

Satellite city

A community that has grown up or been built, frequently around an industrial site within commuting distance of a major city or metropolitan area. See also NEW TOWN.

Savings and loan association (S&L)

A type of cooperative financial institution in the United States that takes in savings as share capital, permitting deposits and withdrawals from interest-bearing, insured savings accounts, and invests the funds in mortgages. The equivalent institution in Great Britain is the building society. See also PRIMARY MORTGAGE MARKET.

"Scenic focal point"

One of the techniques used in traditional Chinese garden design. It aims at the continuous appearance of distant

出售租回 Chūshòu zūhuí

美国的一种交易方式。原业主将其所使用的设备或房地产出售给新业主,然后再从新业主手中租回继续使用一段时间,双方进行这样的安排是为了在财务方面尤其是在减税方面共同获得好处。[参见"避税手段"]

卫星城 Wèixīngchéng

多指环绕某一工业基地建设起来的社区。这些小城与主要城市或大都市的距离一般都在通勤往返的距离之内。[参见"新城"]

储蓄贷款协会 Chǔxù dàikuǎn xiéhuì

美国的一种合作性金融机构,将存款作为股份资本投资于抵押贷款,并允许存款人从其生息和享有担保的储蓄账户上随时进行存取。在英国,类似的机构称为住房协会。[参见"初级抵押贷款市场"]

对景 Duìjǐng

中国古典园林布局手法之一,旨在从某一角度突出地展现一个自然或建筑景观,使观者在步移景

views between the limits of physical elements directly opposite the sight line of the viewer moving through the garden. Similar techniques are adopted in the planning of townscapes; for example, a road may be planned to have the view of a pagoda at its end in the distance. See also "BORROWED VIEW," LANDSCAPE ARCHITECTURE, and "YIJING."

Schedule of values

In the United States, a statement furnished by the contractor to the architect or engineer reflecting the portions of the contract sum allotted for the various parts of the work and used as the basis for reviewing the contractor's applications for progress payments made during the course of work.

Schematic design

In the United States, the initial phase of architectural design in which the architect prepares schematic design studies according to the client's requirements. The documents include single-line drawings showing the general outline of the project and the scale and relationship of project components. These documents, together with an approximate projection of construction costs and a tentative schedule, are submitted to the client for approval. Schematic design is also performed for some projects in China, before

异的过程中，可在园林的实体，如屋顶、窗洞、树木之间，不断欣赏与其视线遥遥相对的此种景观。该手法亦用于城市规划中，例如使街道尽头对着远处一座宝塔等。[参见"借景"、"风景园林学"、"意境"]

工程分项价值表 Gōngchéng fēnxiàng jiàzhíbiǎo

在美国，由承包人致建筑师或工程师的文件，内容包括总金额及各分项金额，是检查承包人按进度申请付款的依据。

方案设计 Fāng'àn shèjì

在美国，指建筑师根据顾客要求而做的初步建筑设计工作。设计文件包括说明工程概貌的单线图、工程各组成部分的尺寸和相互关系，这些设计文件连同建筑成本的概略估计和设想进度表等一起送交委托人，供其批准。在中国，某些工程项目初步设计之前也进行方案设计。[参见"粗略估算"、"设计阶段"]

the preliminary design phase. See also AREA TAKE-OFF and DESIGN PHASE.

Scope of construction project

In China, a standard term used in describing the size of a construction project, as set by the State for various national economic sectors. Projects are usually categorized as "large," "medium," or "small" according to the production capacity or performance efficiency, or to the total amount of investment. For each size class, the State has stipulated different requirements with regard to the design program, the authority responsible for the examination and approval of design program, the authority responsible for the examination and approval of design documents, program arrangements (project start date and all necessary financial, labor, and material resources), and the inspection process for acceptance of completed projects. See also INVESTMENT IN CAPITAL CONSTRUCTION.

Secondary mortgage market

In the United States, a vast financial market in which institutions buy and sell mortgages as investments. After mortgage loans have been issued to property owners by primary lenders in the primary mortgage market, the original lending institutions turn to the

建设项目规模　Jiànshè xiàngmù guīmó

在中国，本词用来反映各类建设项目的规模，分大、中、小三种，由国家根据生产能力或效益或总投资额的大小对国民经济各部门分别规定，以便分级管理。国家对大、中、小型项目的设计任务书的内容、设计文件的审批权限、计划安排（包括工程项目的开工日期以及所需的资金、劳动力和材料）及竣工验收等方面皆有不同要求。[参见"基本建设投资"]

二级抵押贷款市场　Èrjí dǐyā dàikuǎn shìchǎng

在美国，专门以买卖抵押契据为投资经营的巨大金融市场。贷款机构在初级抵押贷款市场上将贷款发放给产业主后，转而在二级市场出售其抵押贷款契据，从中取得资金，并再次发放新的抵押贷款。二级市场可把抵押贷款市场与更大的资金

secondary mortgage market to sell the mortgages in order to obtain the funds necessary to issue new mortgages. This secondary market links the mortgage market to the larger capital market, opening new sources of mortgage credit. Three organizations created by the U.S. Congress to develop the secondary residential mortgage market are the Federal National Mortgage Association (Fannie Mae), the Government National Mortgage Association (Ginnie Mae), and the Federal Home Loan Mortgage Corporation (Freddie Mac). See also MORTGAGE, MORTGAGE BANK, and PRIMARY MORTGAGE MARKET.

Secondary wastewater treatment

The second stage of purifying wastewater, involving removal of dissolved organic contaminants from sewage through the activated sludge process or through use of the trickling filter or some type of lagoon. See also ADVANCED WASTEWATER TREATMENT, PRIMARY WASTEWATER TREATMENT, SEWAGE DISPOSAL SYSTEM, WASTE MANAGEMENT, and WASTE UTILIZATION.

Sector planning

In the United States, planning of metropolitan land

市场联系起来,开创新的抵押信贷资源。美国国会为发展住房抵押贷款二级市场创建了三个组织:联邦全国抵押贷款协会、政府全国抵押贷款协会、联邦住房抵押贷款公司。[参见"抵押贷款"、"抵押贷款银行"、"初级抵押贷款市场"]

污水二级处理　Wūshuǐ èrjí chǔlǐ

净化污水的二级处理,包括采用活性污泥法、滴滤池或某些类型的生物塘,除去水中的溶解性有机污染物质。[参见"污水深度处理"、"污水一级处理"、"污水处理系统"、"废物管理"、"废物利用"]

局部规划　Júbù guīhuà

在美国,将大都市地区分为相互独立的区或部分,

use and housing production by dividing the urban area into mutually exclusive zones or sectors and determining supply and demand on a sector-by-sector basis.

Sectorization of functions

For planning purposes in China, division of a city into various districts according to their functions, e.g., commercial, educational, and industrial districts. See also DISTRICT.

Security

1. Property of some sort, e.g., stocks or land, held as a guarantee of loan repayment. See also LIEN.

2. Precautionary measures taken to prevent illegal entry into a building or property. These can be architectural, technological, or human (e.g., guards).

Segregated use

In the United States, differentiation of urban land into special purpose areas, for example, residential or commercial districts, as the result of economic forces, social attitudes, or zoning regulations.

并确定每部分内的供求关系，从而做出土地使用及住房建造的规划。

功能分区 Gōngnéng fēnqū

在中国，指规划工作中，按不同使用功能将城市划分为若干区，如商业区、文教区、工业区等。[参见"区"]

1. **担保品** Dānbǎopǐn

作为贷款保证的资产，如土地或股票等。[参见"留置权"]

2. **安全措施** Ānquán cuòshī

为防止他人非法进入建筑物或产业而采取的安全措施。可以是建筑上的、技术方法上的措施或采用人的因素，如雇用警卫等。

分区专用 Fēnqū zhuānyòng

在美国，将都市土地分成不同用途的地区。例如，根据经济实力、社会倾向或区划规定而分成住宅区和商业区等。

Seismic planning

In China, a constituent part of comprehensive planning which includes determination of the earthquake scale to be used and the earthquake intensity that designed structures should be able to withstand; differentiation of land areas according to their suitability for construction, as well as limitations on form, height, and number of stories of buildings; provision of evacuation facilities; utilization of underground structures; prevention of secondary damage from disasters; and recommendation of remedial measures. See also DISASTER PLANNING.

Self-help housing

Dwellings designed and constructed partially or completely by the individual owners themselves to reduce costs. See also CORE HOUSING, HOME IMPROVEMENT, and OWNER-BUILDER.

Semidetached house

See DUPLEX HOUSE.

Service facility

In China, a general term for any facility providing an essential service, including schools, stores, repair shops, clinics, and so forth. For the U.S. equivalent, see also COMMUNITY FACILITY.

城市抗震规划 Chéngshì kàngzhèn guīhuà

在中国，是城市总体规划的组成部分，内容包括确定城市建设的抗震等级、工程设计的抗震烈度，从地震角度考虑划分可建和不可建用地、建筑物高度、体形与层数的限制，疏散通道、地下建筑的利用，次生灾害的防止及补救措施等。［参见"城市防灾规划"］

自建住房 Zìjiàn zhùfáng

为了节约造价而全部或部分由私人自己设计和建造的住宅。［参见"核心住房"、"住房改善"、"自建业主"］

半独立式住宅 Bàndúlìshì zhùzhái

［见"二联式住宅"］

生活服务设施 Shēnghuó fúwù shèshī

在中国，指为居民生活所需的学校、商店、修理站、卫生所等项目的统称。［美国的相应词汇参见"社区设施"］

Service fee norm

In China, the standard fixed by the responsible authorities for calculating miscellaneous direct expenses, indirect expenses, and extra expenses, which are to be charged according to certain percentage rates of the direct expenses, basic wages, or construction costs. Expenses incurred on construction projects for labor, material, and equipment can be calculated on the basis of norms for detailed estimates. See also NORM.

Service floor area

In China, the total net floor area, apart from living floor area, in a dwelling unit, including the kitchen, the bathroom, and closets. See also RESIDENTIAL FLOOR AREA.

Service population

In China, the part of the labor population which, in contrast to the primary population, provides only services to the local community in the cultural, administrative, and commercial sectors. See also LABOR POPULATION and PRIMARY POPULATION.

Service sector

Businesses in the private-sector economy that

取费标准　Qǔfèi biāozhǔn

在中国，指由主管部门制订的关于各种直接费、施工管理费和独立费等的费用定额，系按照直接费、基本工资或工程成本的一定百分比计取。直接用于工程的人工费、材料费、机械费则可根据预算定额直接计算。[参见"定额"]

辅助面积　Fǔzhù miànjī

在中国，指一个居住单元内除居住面积外的全部建筑净面积，包括厨房、浴室、壁橱等。[参见"居住建筑面积"]

服务人口　Fúwù rénkǒu

在中国，指劳动人口中除去基本人口以外的部分，即城市中为当地服务的行政机关工作人员和文化、商业服务机构中的职工。[参见"劳动人口"、"基本人口"]

服务行业　Fúwù hángyè

指专门提供服务的行业，例如美国的旅游业、保

specialize in provision of services rather than in the manufacture and sale of goods to commercial enterprises and the public, for example, in the United States, travel and insurance agents, and in China, mainly hotels, barbershops, laundries, and professional photographic studios.

Serviced land

Land which has been provided with infrastructure facilities, such as access roads, sewers, water mains, and power lines. See also INFRASTRUCTURE and SITE DEVELOPMENT.

Setback

In the United States, placement of a structure some distance from the front, side, or back lot line of the building site to improve the appearance of low buildings on large lots, to prevent overshadowing of streets and adjacent structures by high-rise buildings, or to allow maximum yard space or distance between houses. Setback requirements are frequently defined by zoning ordinances.

Settlement costs

In the United States, the charges and fees which must be paid, usually by the purchaser of property, at the time when the official documents for the purchase

险业和中国的旅馆业、理发店、洗衣店和照相馆等。他们并不制造货物，也不向商业企业或公众销售货物。

有设施用地　Yǒushèshī yòngdì

设有出入通道、上下水管道及电源线路等基础设施的土地。[参见"基础设施"、"建设场地开发"]

缩进　Suōjìn

在美国，指建筑物的四周与其基地边线之间保持一定距离的措施，用以改善建于较大地块上的低层建筑物的外观，避免建筑物和街道被邻近的高层建筑物遮挡，在住宅之间提供最大的庭院空间或距离。对于缩进的要求，一般在区划法令中予以规定。

成交费用　Chéngjiāo fèiyòng

在美国，指房地产买卖成交签订正式文件时，通常需由买方支付的费用，包括律师费、产权证保险费、产权检验费、估价和检查费、贷款手续费、

transaction are signed. The costs include lawyers' fees, a title insurance premium, a title examination fee, appraisal and inspection fees, loan charges, and state taxes and fees. These costs vary considerably from state to state. Also called closing costs.

Sewage disposal system

Complex of equipment and processes designed to render raw sewage harmless through treatment at the individual home or community level and to dispose of the treated sludge and wastewater components in a nonpolluting, useful manner. See also ADVANCED WASTEWATER TREATMENT, PRIMARY WASTEWATER TREATMENT, SECONDARY WASTEWATER TREATMENT, WASTE MANAGEMENT, and WASTE UTILIZATION.

Shared appreciation mortgage

In the United States, a mortgage agreement in which the borrower agrees to transfer a percentage of a property's appreciation in value to the lender, usually at the time of sale, in exchange for a stable, below-market interest rate or a small downpayment requirement. There are many variations on this arrangement under the general name of shared equity plans. See also APPRECIATION and MORTGAGE.

州定税收等。这些费用随州而变,往往差异很大,又称"结算费用"。

污水处理系统　Wūshuǐ chǔlǐ xìtǒng

使家庭或社区的污水变为无害的一整套设备和过程,同时将经过处理的泥浆和废水以不造成污染、变废为利的形式排放出去。[参见"污水深度处理"、"污水一级处理"、"污水二级处理"、"废物管理"、"废物利用"]

分享增值抵押贷款　Fēnxiǎng zēngzhí dǐyā dàikuǎn

在美国,一种抵押贷款的做法。借款者同意将房地产增值的一定百分比转让给贷款者,以换取低于市场利率的稳定贷款或较小的首次付款要求。这种转让常在出售房地产时进行,分享增值的做法名目繁多,但总的都称为分享资产净值计划。[参见"增值"、"抵押贷款"]

Shared occupancy

In China, more than one officially registered household living in a single dwelling. See also HOUSING SHORTAGE, OCCUPANCY, and SINGLE OCCUPANCY.

Shelter

1. Any kind of structure used for residential purposes.
2. A temporary or emergency living facility to house disaster victims or the homeless. See also EMERGENCY HOUSING.

Shopping center

See SHOPPING MALL.

Shopping mall

In the United States, a large, often enclosed shopping area, usually one or two stories high, containing a variety of retail and department stores along an interior pedestrian walk. Malls are generally located in the suburbs and are surrounded by parking areas. Previously, such collections of retail stores were seldom enclosed and were (and still are) referred to as shopping centers.

Side effect

An anticipated or unanticipated, positive or negative secondary consequence of a policy, process, project,

合住 Hézhù

在中国，指一套住房由几户合住。[参见"住房短缺"、"占用"、"独门独户"]

栖身处 Qīshēnchù

1. 任何可供居住的建筑物。
2. 为灾民或无家可归者提供的临时或紧急的居住设施。[参见"应急住房"]

购物中心 Gòuwù zhōngxīn

[见"购物商场"]

购物商场 Gòuwù shāngchǎng

在美国，指大的封闭型购物区，多为一至两层，常位于郊区，四周有停车场。商店包括各种零售店和百货公司，沿内部人行走道设置。以前这种零售商店集中的地段很少采取封闭型，称"购物中心"，现在也仍有称"购物中心"的。

副作用 Fùzuòyòng

某一政策、进程、工程项目或行动所造成的预期的或非预期的、正面的或反面的间接结果。[参见

or action. See also EXTERNALITIES and IMPACT.

Sideline production

In rural areas of China, production of goods not directly related to the regular agricultural, forestry, animal husbandry, and fishing activities. Such products, e.g., handicraft items, are produced in farmers' spare time and are sold in State-owned shops or in the free market. See also COTTAGE INDUSTRY.

Single-family detached house

See DETACHED HOUSE.

Single-generation household

In China, a household consisting of persons without children or not living with their children. See also HOUSEHOLD.

Single-occupancy

Exclusive use of a dwelling by a single person or family. See also OCCUPANCY and SHARED OCCUPANCY.

Sinking fund

In the United States, an account established to receive regular, equal deposits that are accumulated to pay off a loan or debt when it falls due. Such accounts are usually

"外在副作用"、"影响"]

副业生产 Fùyè shēngchǎn

在中国,指农村中除了农业、林业、畜牧业、渔业以外的生产,其产品为农民在农闲时以手工制作并在国营商店或自由市场出售。[参见"家庭工业"]

独立式独户住宅 Dúlìshì dúhù zhùzhái

[见"独立式住宅"]

无子女户 Wúzǐnǚhù

在中国,指无子女或不与子女同住的户。[参见"户"]

独门独户 Dúmén dúhù

一套住房由一人或一户独住。[参见"占用"、"合住"]

偿债基金 Chángzhài jījīn

在美国,为积累款项供到期偿清贷款或债务而开立的定期、定额存款户头。此种户头通常可生利息,贷款机构根据账面存款金额付给储户利息。

interest-bearing, i.e., the lending institution pays the depositor interest on the money deposited in the account.

Site analysis

Careful examination of a land parcel prior to construction to ensure that all site-related functional, structural, and ecological development proposals are properly coordinated with inherent site capacities and that potential problems associated with distinctive features of the site are anticipated and resolved.

Site architect

See RESIDENT ARCHITECT.

Site development

The division and excavation of a piece of land and its improvement through provision of an infrastructure and through construction. See also SERVICED LAND and SITE ENGINEERING.

Site engineer

See RESIDENT ENGINEER.

Site engineering

Planning of a development area, taking into consideration the characteristics and functional requirements of the construction project, as well as the

建设场地分析 Jiànshè chǎngdì fēnxī

施工之前所进行的场地检查。以确保与场地的功能、结构、生态发展等有关的建议能与场地的内在能力相适应，并能预见到与该场地明显特征有关的潜在问题，事先予以解决。

现场建筑师 Xiànchǎng jiànzhùshī

［见"驻场建筑师"］

建设场地开发 Jiànshè chǎngdì kāifā

对一块土地的划分和开挖，并通过提供基础设施和施工，使该地块得到改善。［参见"有设施用地"、"竖向规划"］

现场工程师 Xiànchǎng gōngchéngshī

［见"驻场工程师"］

竖向规划 Shùxiàng guīhuà

对建设场地，按其自然状况、工程特点和使用要求所做的规划。包括：场地与道路标高的设计，建筑物室内、外地坪的高差等，以便在尽少改变

natural features of the site. Site engineering considers the levels of the site and the roads, and the difference between the levels of the ground and of the ground floor. The objective is to achieve a site that fills the needs of the intended occupant and provides the basis for good drainage and durable construction with a minimum of disturbance to the existing topography and natural amenities. See also SITE DEVELOPMENT.

Site inspection

Examination of a structure on the construction site to ensure that the work conforms to requirements or specifications. See also BUILDING INSPECTOR, CONSTRUCTION INSPECTOR, and NONCONFORMING WORK.

Site planning

The design and preparation of plans, specifications, and construction details for a land parcel under development, taking into consideration such details as placement of facilities, grading, drainage, utilities, roads, walkway, parking, planting, construction details, and so forth.

Sites and services

A development technique which involves providing low-cost lots with basic infrastructure services (e.g.,

原有地形及自然景色的情况下满足日后居住者的要求,并为良好的排水条件和坚固耐久的建筑物提供基础。[参见"建设场地开发"]

现场监督 Xiànchǎng jiāndū

在施工现场对建筑物的检查,以保证工程质量符合规程的要求。[参见"建筑监察员"、"施工检查员"、"不符合同工程"]

建设场地规划 Jiànshè chǎngdì guīhuà

为某一地块的建设所准备的设计和平面布置图、说明书及工程细节,包括对设施的位置、标高、排水、市政设施、道路、人行道、停车场、绿化等细节的考虑。

提供场地与基础设施 Tígōng chǎngdì yǔ jīchǔ shèshī

为低收入城市居民在低价土地上提供基本的基础设施(如公共厕所和共用水源)的一种开发技术。

latrines and community water sources) to low-income urban residents, particularly in developing countries. See also CORE HOUSING.

Size distribution of dwellings

In China, the percentage of each type of housing unit, classified by the number of habitable rooms, as compared to the total number of housing units in a housing project. See also DWELLING SIZE.

Size distribution of households

The number and percentage of households with one, two, three, four, or more persons. Such statistics may also include information on the age and sex distribution of households. See also HOUSING SURVEY.

Skyline

The outline of a configuration of tall buildings against the sky, a characteristic feature of large, modern cities such as New York. See also CITYSCAPE.

S&L

See SAVINGS AND LOAN ASSOCIATION.

Slum

A predominantly residential urban area characterized

特别应用于发展中国家。[参见"核心住房"]

户室比 Hùshìbǐ

在中国,指不同居室数的各类居住单元各占居住单元总数的百分比。[参见"户型"]

户型比 Hùxíngbǐ

不同人口数的各类户数及其在总户数中所占的百分比。这种统计也可包括年龄、性别的资料。[参见"住房调查"]

天际线 Tiānjìxiàn

由许多高大的建筑物在天际形成的轮廓,为现代化大城市(如纽约)的一个特色。[参见"城市风貌"]

储蓄贷款协会(缩写) Chǔxù dàikuǎn xiéhuì(suōxiě)

[见"储蓄贷款协会"]

贫民窟 Pínmínkū

都市中住房破旧、缺乏管理、街道脏乱的贫困居

by deteriorating housing poorly kept, littered streets, and poverty. In the United States, urban redevelopment programs frequently involve slum clearance, i.e., demolition or improvement of slum dwellings to provide acceptable housing. See also BLIGHT, TENEMENT BUILDING, and URBAN REDEVELOPMENT.

Small business

As defined by the U.S. Small Business Act of 1953, a commercial undertaking which is independently owned and operated, and which is not a dominant force in its area of activity. The specific number of employees and the volume of business used to define a small business vary from industry to industry.

Small city

See CITY SIZE.

SMSA

See METROPOLITAN STATISTICAL AREA.

Social indicator

A measure or data type regarded to be predictive of social performance or behavior within a human population group. The publication "Social Indicators",

住区。在美国,城市改造计划常涉及贫民窟的清除。例如,拆除或改善贫民窟的住所以提供合理的住房。[参见"衰落"、"经济公寓"、"城市改造"]

小型企业 Xiǎoxíng qǐyè

按照美国1953年《小型企业法》的定义,指独自拥有和经营的商业机构。小型企业在其本行业活动领域中不是主导力量。关于小型企业雇用人员数和营业额的规定,随不同行业而异。

小城市 Xiǎochéngshì

[见"城市规模"]

标准大都市统计区(缩写) Biāozhǔn dàdūshì tǒngjìqū (suōxiě)

[见"大都市统计区"]

社会指数 Shèhuì zhǐshù

用以预测某一人口集团的社会行动或行为的一种方法或数据资料。由美国商业部定期发表的社会指数包括诸如犯罪率、结婚率、离婚率、收入水

produced periodically by the U.S. Department of Commerce, contains statistics on social conditions and trends such as crime rates, marriage and divorce rates, and income levels.

Social integration

The merging of divergent social groups into the mainstream of a society through their adjustment to generally accepted social standards and behaviors.

Social participation

See CITIZEN PARTICIPATION.

Social planning

Efforts to anticipate and resolve problems that arise in the social environment of a community or organization. In the United States, government social planning is restricted to narrowly defined areas of social concern, e.g., provision of health or welfare services, at the local city or district level.

Social services

See HUMAN SERVICES.

Social structure

The overall organization of the interrelationships among individuals, organizations, and institutions of a society,

平等社会情况和发展趋势。

社会一体化 Shèhuì yītǐhuà

不同的社会集团按一般均能接受的社会准则和行为自行调整，从而融汇于社会的主流之内。

社会参与 Shèhuì cānyù

［见"市民参与"］

社会规划 Shèhuì guīhuà

预测和解决在一个社区或组织的社会环境中所产生的问题。在美国，政府的社会规划仅限于极狭窄的与社会有关的领域内，如在地方城市或市区一级提供保健或福利服务。

社会服务 Shèhuì fúwù

［见"人类服务"］

社会结构 Shèhuì jiégòu

指一个社会中个人、组织和社会部门之间的相互关系，包括文化价值与行为、阶级与财富的分配、人

with consideration of cultural values and behaviors, class, distribution of wealth, demography, and a variety of other factors. See also STRATIFICATION.

Socio-economic mix

The constitution of the population of an urban area or neighborhood, as differentiated by social factors such as cultural background, race, and religion and by economic factors such as income level and professional status.

Soil survey

Study and chemical analysis of the soil characteristics in a particular area, including classification and mapping of properties, identification of soil types, and determination of soil suitability for purposes such as construction and cultivation.

Space planning

Design of the use areas within a building to ensure provision of space adequate to meet human living, comfort, and recreation requirements. See also INTERIOR DESIGN.

Special district

In the United States, a zone with designated uses to meet special requirements of certain areas, e.g.,

口学以及其他多种因素。[参见"社会阶层划分"]

社会经济综合指标　Shèhuì jīngjì zōnghé zhǐbiāo

在美国,指按文化背景、种族与宗教等社会因素以及按收入水平与职业地位等经济因素,区分一个城市地区邻里的人口构成。

土壤调查　Tǔrǎng diàochá

对某地区内土壤特性的研究与化学分析,包括对地产的分类和制图、确定土壤类型及其对于建设或耕种的适用性。

空间布局　Kōngjiān bùjú

为提供满足人们居住、舒适和娱乐要求的空间而对建筑物内部的面积使用所做的设计工作。[参见"室内设计"]

特区　Tèqū

在美国,被指定作特殊用途的特定地区。例如在飞机场附近的建筑物有高度限制,在泛滥平原禁

limitation of building heights around airports or of residential construction on floodplains, or restriction of use to farming. See also DISTRICT and ZONING DISTRICT.

Special economic zone

In China, a region granted special permission of the State to operate its economy and administration according to relatively independent domestic and foreign policies.

Special use permit

In the United States, written approval of the planning authority allowing land in a zoned area to be used for unusual purposes extending beyond normal district uses or for uses as specified in a special or conditional use list contained in the relevant zoning ordinance. Approval is granted only after a report by the planning director and a public hearing. See also BOARD OF ZONING APPEALS and ZONING VARIANCE.

Specialized team

See CONSTRUCTION TEAM.

Specification code

A building code establishing construction requirements

止建造住宅，或某些地区禁止用于农业等。[参见"区"、"区划市区"]

经济特区 Jīngjì tèqū

在中国，指由国家特许，在经济上和行政管理上实行相对独立的对内与对外政策的地区。

专用许可证 Zhuānyòng xǔkězhèng

在美国，指规划当局书面批准的特别许可证，允许将区划地区内的土地用于超出一般用途的特殊目的，或用于有关的区划法令规定的特殊的或有条件的使用范围，必须由规划主管人写出报告并经过公众意见听取会后才能批准此种专用许可证。[参见"区划申诉委员会"、"区划特许证"]

专业施工队 Zhuānyè shīgōngduì

[见"施工队"]

规格法规 Guīgé fǎguī

主要参照特定的施工方法和材料而制定施工要求的

predominantly by reference to particular construction methods and materials, as distinguished from a performance-oriented code. See also BUILDING CODES AND STANDARDS, PERFORMANCE CODE, and SPECIFICATIONS.

Specifications

1. Requirements for materials and construction prescribed by building codes. See also BUILDING CODES AND STANDARDS and SPECIFICATION CODE.
2. A detailed description of a proposed structure provided by the architect to the owner.
3. The part of the contract documents for the construction of a building that includes requirements for materials, equipment, construction systems, standards, and workmanship. See also CONTRACT DOCUMENTS.

Speculation

In the United States, purchase of land, property, or securities at a low price with the intention of selling in the short or long term at a considerable profit.

Split-level house

In the United States, a house with several adjacent living areas at different elevations. Often on sloping ground and with entrances at different levels.

建筑规范，以区别于以功能为准则的法规。[参见"建筑法规与标准"、"功能法规"、"建筑构造规格"]

1. **建筑构造规格** Jiànzhù gòuzào guīgé
建筑法规中所规定的材料和施工要求。[参见"建筑法规与标准"、"规格法规"]
2. **建筑细部说明** Jiànzhù xìbù shuōmíng
建筑师向业主提交的拟议中的建筑物的详细说明。
3. **说明书** Shuōmíngshū
工程施工合同文件的一部分，其中包括对材料、设备、施工体系、标准及施工质量的要求。[参见"合同文件"]

投机 Tóujī
在美国，指以低价购进土地、产业或债券，然后在可获利的时机出售，从中获利。

错台式住宅 Cuòtáishì zhùzhái
在美国，指几个部分分别位于不同标高而又相连的住宅。通常建于斜坡上，各标高分别有其进出口。

Spot zoning

In the United States, zoning for a single property that is different from that of properties in the surrounding area; an approved exception to the general zoning of an area. See also ZONING.

Squatter settlement

A collection of nonconventional housing or shanty towns spontaneously constructed by the urban poor, often on land not owned by the individuals constructing the dwellings. Such settlements are particularly common on the fringes of cities in Third World countries.

"Squeeze-in development"

In China, inappropriate development of a densely settled older city by constructing buildings in any available space between existing structures. See also INFILL DEVELOPMENT and PIECEMEAL DEVELOPMENT.

Standard Metropolitan Statistical Area

See METROPOLITAN STATISTICAL AREA.

Standard mortgage

See FIXED-RATE MORTGAGE.

State government

In the United States and countries with a comparable

个别点区划 Gèbiédiǎn qūhuà

在美国，指对不同于周围地产的个别地产所做的区划，亦即在一个一般区域地区内获准的例外。[参见"区划"]

违章居留地 Wéizhāng jūliúdì

在非自有的土地上由城市贫民自发地草率搭建的非正式住房或棚户区。在第三世界的城市边缘尤为普遍。

"见缝插针" Jiànfèng chāzhēn

在中国，指在人口稠密的旧城建设中，不恰当地在已有的建筑物之间的空地上进行建筑。[参见"填空性建设"、"零星建设"]

标准大都市统计区 Biāozhǔn dàdūshì tǒngjìqū

[见"大都市统计区"]

标准抵押贷款 Biāozhǔn dǐyā dàikuǎn

[见"固定利率抵押贷款"]

州政府 Zhōuzhèngfǔ

在美国和类似联邦制的国家，在州一级被授权制

federal structure, the authority empowered to make public policy decisions, to enact law, and to carry out public administrative functions at the state level. State government falls in the administrative hierarchy below the federal government and above local government. See also LOCAL GOVERNMENT.

State investment

In China, a principal source of financial resources for construction projects. The amount invested is listed in the State investment plan for fixed assets, and funds are allocated through State budgetary appropriations. See also INVESTMENT IN CAPITAL CONSTRUCTION.

Statute of limitations

In the United States, a regulation establishing the length of the time period in which legal action can be brought for damages or relief under provisions of a given law.

Statutory profit

In China, the allowable profit for contractors, calculated at a rate set by financial authorities according to the estimated cost of work already completed on the particular project. See also

定公共决策、拟订法律、行使公共管理职能的领导部门。在管理层次上，州政府在联邦政府之下、地方政府之上。[参见"地方政府"]

国家投资 Guójiā tóuzī

中国建设项目资金来源之一，指列入国家固定资产投资计划并由国家财政预算拨款的投资。[参见"基本建设投资"]

时效 Shíxiào

美国的一种法律规定，即当根据某法律条款要求赔偿损害或予以减免时，可采取法律行动的期限。

法定利润 Fǎdìng lìrùn

在中国，指施工单位根据国家规定的利润率按已完工程的预算成本所提取的利润。[参见"建筑造价"]

CONSTRUTION PAYMENT.

"Stick-built"

In American construction slang, a house constructed on site with a wooden frame. The term is usually applied to single-family residential housing. This type of housing represents the opposite pole to manufactured housing, even though many components of a "stick-built" house are produced in a factory.

Stratification

In sociology, hierarchical division of individuals or groups of individuals into strata or classes according to a defined characteristic or set of characteristics. See also SOCIAL STRUCTURE.

Street furniture

A collective term designating benches, signs, lights, fixtures, trash receptacles, and other items provided along a street right-of-way.

Streetscape

The view along a street created by the buildings, green spaces, and street furniture. See also CITYSCAPE.

木架房屋　Mùjià fángwū

在美国建筑业的俚语中,指在现场建造的木构架房屋,一般指独户住宅。虽然木架房屋的许多部件是由工厂制造的,但这种住房代表与工厂预制住房相反的类型。

社会阶层划分　Shèhuì jiēcéng huàfēn

在社会学中,根据已确定的某一项或某一组特征,将个人或团体划分成不同阶层或阶级。[参见"社会结构"]

街道小品　Jiēdào xiǎopǐn

街道两边各项设施的统称,如座椅、路标、路灯、垃圾筒等。

街景　Jiējǐng

由沿街建筑群与绿化、小品等形成的综合景观。[参见"城市风貌"]

Sculptural Style
Dulles Airport, 1962
near Washington, D.C.
(Architects: Eero Saarinen and Associates)

杜勒斯国际机场

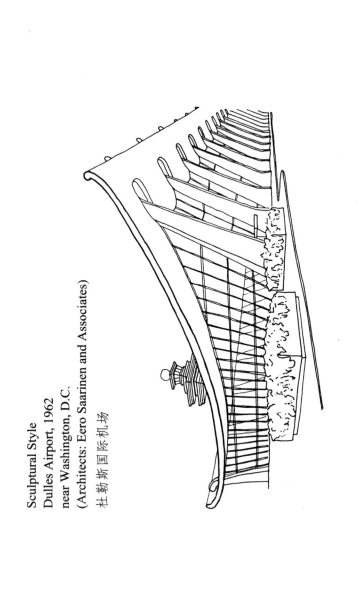

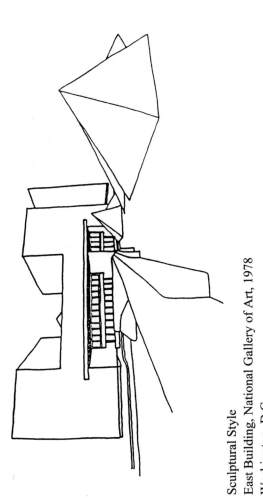

Sculptural Style
East Building, National Gallery of Art, 1978
Washington, D.C.
(Architect: I. M. Pei)

贝聿铭建筑师设计的国际美术东馆

Strip development

In the United States, a narrow section of land along a major street or highway that is zoned for commercial use. Common earlier in small American cities and towns, this type of development is now discouraged because of a variety of side effects: numerous, unattractive neon signs of different sizes, problems of traffic safety on main streets, and the unsightly back sides of businesses abutting directly on residential areas. Also called a ribbon development.

Structural unemployment

Joblessness for certain types of workers, resulting from changes in the structure of consumer demand and technology and from consequent alterations in the demand for labor without corresponding changes in the composition of the labor force. Often such unemployment is also geographic, with a surplus of particular types of workers in a region of declining industry and a short supply in expanding areas. See also OCCUPATIONAL STRUCTURE.

Studio apartment

Traditionally, a living unit that has a large, well-lighted room with a high ceiling suited for artistic endeavors, such as painting or sculpting. The term is now applied to any one-room apartment with a

条形商业区　Tiáoxíng shāngyèqū

在美国，沿公路或主要街道划为商业用途的狭长地段。美国早期的市镇中多有这种商业区，但目前不鼓励这种建造方式，因为它有很多副作用，如大量参差不齐的霓虹灯、主要街道上的交通安全问题以及商业街背后不雅的外观直接贴近住宅区等，亦称"带形商业区"。

结构性失业　Jiégòuxìng shīyè

由于消费结构和技术的改变，带来对劳动力需求的改变，但因未在劳动力构成上做相应的改变，而导致某些工种的工人失业。这种失业经常是地理性的，即在某一区域因工业的衰退而造成某些工种的工人过剩，而在另一发展地区又有不足。
［参见"职业结构"］

画室公寓　Huàshì gōngyù

传统上，指具有一间高大明亮的适合艺术家从事绘画或雕刻等创作的房间的居住单元。目前则指任何一种带有小厨房和浴室的一室户公寓。

kitchenette and bathroom.

Subcontract

An agreement between the prime contractor and another contractor or a supplier for specific services, materials, or a stipulated portion of the work, as indicated on the plans and specifications, all as evidenced by the contract documents. See also CONTRACT and PRIME CONTRACT.

Subcontractor

See CONTRACTOR.

Subdistrict office

In China, the grass roots unit of government administration in a subdivision of the city district. The subdistrict office is the agency of the city district government responsible for directing the neighborhood committees and overseeing neighborhood industry and public affairs. See also DISTRICT.

Subdivision

In the United States, a piece of land that has been surveyed and divided into lots for residential use. The term is usually applied to developments constructed in suburban areas after World War II. See also NEIGHBORHOOD.

分包合同　Fēnbāo hétóng

在美国，由直接与业主签订合同的承包人和另一承包人或供应商之间所订立的合同，规定按原合同文件的图纸和说明书提供符合要求的服务或材料，或承担部分工程。[参见"合同"、"直接承包合同"]

分包人　Fēnbāorén

[见"承包人"]

街道办事处　Jiēdào bànshìchù

在中国，指街区范围内的基层政权机构，是区政府下属的派出单位，领导所属街区内的居委会，主办街道工厂，处理公众事务。[参见"区"]

分块土地　Fēnkuài tǔdì

在美国，指一块土地经测量后被划分成若干小块，作居住用。此词多用于第二次世界大战后的城市郊区建设。[参见"邻里"]

Subsidized housing

See ASSISTED HOUSING.

Subsidy

Any kind of grant or aid extended by the government to an undertaking which serves the public interest, such as a housing subsidy. See also HOUSING ALLOWANCE.

Substandard housing

Dwellings that are poorly maintained and inadequately equipped, endangering the health and safety standards of occupants. See also DILAPIDATED HOUSING and HOUSING SHORTAGE.

Suburb

The outlying section of a city or a smaller adjacent town which has historically served primarily as a residential area. In the United States in recent years, suburbs have become more diversified with industrial and commercial as well as residential uses. In China, the portion of a city's suburbs closest to the city proper is officially designated the inner suburban district, while the outlying suburbs are called the outer suburban district. The outer suburban district and the municipally administered counties are in the same geographic belt. See also COUNTY.

补贴住房 Bǔtiē zhùfáng

[见"公助住房"]

补贴 Bǔtiē

由政府为公众利益而授予的赠款或补助,如住房补贴。[参见"住房津贴"]

不合标准住房 Bùhébiāozhǔn zhùfáng

指维修不善、设备不足、危及住户健康与安全的住宅。[参见"危房"、"住房短缺"]

郊区 Jiāoqū

城市的边缘地带或附近的小城镇,历史上主要为住宅区。近年来,美国的城市郊区已渐趋多元化,有工业、商业和居住等多种用途。在中国,离城市近的郊区由官方正式确定为近郊区,而近郊区以外的市属县所在地称远郊区。[参见"县"]

Sunlight ordinance

A regulation protecting the right of property owners to have sunlight falling continuously on their property. Such ordinances are rare in the United States but are common and highly refined in Japan. In China, ordinances define the minimum distance between two rows of parallel buildings, calculated according to the designated minimum duration of sunlight shining through the windows of the back row of buildings on the day of the winter solstice, without being obscured by the front row. See also BUILDING CODES AND STANDARDS.

Support item

In a planned development of a residential district or housing estate in China, all the structures except the main ones, e.g., the apartment buildings, are referred to as support items, including the service facilities and public utilities.

Sweat equity

In the United States, the money value of a property in excess of the money owed on that property, resulting from the labor invested by the individual owner to improve the property. For low-income families, sweat equity can substitute for a downpayment in obtaining basic housing. See also EQUITY and HOME

日照标准 Rìzhào biāozhǔn

保护业主的产业能经常得到日照的一种规定。日本对日照间距有详细规定,在美国则很少见。中国规定:两排互相平行的建筑物间允许的最小间距,以后排建筑物在冬至日能由窗子射入室内的日照不得少于规定的持续时间为准。[参见"建筑法规标准"]

配套项目 Pèitào xiàngmù

在中国,根据规划建设的居住区或居住小区中,除住宅外用于服务与公用设施的建筑物和构筑物,统称为配套项目。

改进后净增值 Gǎijìnhòu jìngzēngzhí

在美国,指由于业主投入人力改进房地产而使该产业价值提高后超过其负债部分的金额。低收入家庭可用此项资产净增值作为获取基本的住房的首次付款。[参见"资产净值"、"住房改善"]

IMPROVEMENT.

Systems analysis

Use of mathematical techniques to define objectives for a business, government, or military system; to define alternatives in terms of cost and effectiveness for attainment of the objectives; and then to reassess the objectives and develop new alternatives in a continuous cycle. See also SYSTEMS APPROACH and SYSTEMS MANAGEMENT.

Systems approach

A process for investigation and problem-solving based on the application of scientific methods of research, experimentation, and logical analysis. Instead of fragmented, compartmentalized analysis, emphasis is placed on a holistic view of complex organizations, focusing on interrelations and interactions among parts from which emerge the properties of the whole. See also SYSTEMS ANALYSIS and SYSTEMS MANAGEMENT.

Systems management

The planning and direction of a business or government organization or activity using such techniques as systems analysis. See also SYSTEMS ANALYSIS and SYSTEMS APPROACH.

系统分析 Xìtǒng fēnxī

使用数学方法确定商业、政府或军事系统的活动目标，以及为达到此目标可采取的各种替换方案的成本和效益。然后用这种方法反复评定目标，并发展新的替换方案。[参见"系统方法"、"系统管理"]

系统方法 Xìtǒng fāngfǎ

运用研究、实验与逻辑分析等科学方法，进行调查与解决问题的一种步骤。重点放在对复杂事物的全面观点，着重研究决定事物整体性质的各部分之间的内在联系与相互作用，而不是作片段、分隔的分析。[参见"系统分析"、"系统管理"]

系统管理 Xìtǒng guǎnlǐ

采用系统分析技术，对商业和政府机构及其各种活动制订计划并进行指导。[参见"系统分析"、"系统方法"]

T

Task force

In the United States, a group of experts appointed by the government to investigate a particular issue or problem of special concern to the general public and to report its findings and recommendations for improvement of the existing situation.

Tax shelter

In the United States, a legal investment device used by citizens to reduce personal income taxes by taking advantage of tax benefits provided under current tax laws. Common forms of tax shelters are tax-exempt revenue bonds and sale-leaseback arrangements. Abusive tax shelters involve schemes to generate artificial deductions from taxable gross income without real risk or loss to the taxpayer. See also SALE-LEASEBACK.

TDR

See TRANSFERABLE DEVELOPMENT RIGHTS.

Technical assessment

See CERTIFICATION OF MATERIALS AND PRODUCTS.

Technology assessment

Evaluation of the applications of scientific knowledge

特别工作组 Tèbié gōngzuòzǔ

在美国，指由政府指派的专家组，针对与公众有关的某一问题作深入调查，并为改进现状提出报告和建议。

避税手段 Bìshuì shǒuduàn

在美国，公民为减免个人所得税，利用现行税收法中有利于减税的规定所采取的合法的投资策略。常用的避税手段有免税政府债券、出售租回等。纳税人在并无真正风险或损失的情况下故意减报应纳税的总收入则属滥用避税手段。[参见"出售租回"]

可转让开发权（缩写） Kězhuǎnràng kāifāquán (suōxiě)

[见"可转让开发权"]

技术性鉴定 Jìshùxìng jiàndìng

[见"材料与制品鉴定"]

技术评定 Jìshù píngdìng

对科学知识的应用及应用此种知识所生产的产品

and of products resulting from such applications to determine their effectiveness, usefulness, impact, and safety. See also CERTIFICATION OF AMTERIALS AND PRODUCTS and QUALITY CONTROL.

Technology transfer

The useful and appropriate application of technology from one field, economic sector, or country to another.

Tenant services

Special activities performed on behalf of renters or leaseholders of property by the owner or employees of the owner, for example, maintenance work.

Tenement building

In the United States, a building containing a number of dwellings. Originally used to designate the early apartment houses in older American cities, the term now has a negative connotation, as the buildings and the urban areas in which they are located tend to be run down and obsolete. See also APARTMENT, APARTMENT HOUSE, DILAPIDATED HOUSING, and SLUM.

Ten-thousand-yuan norm

In China, a norm for material requirements that

进行评价,以确定其有效性、有用性、影响和安全性能。[参见"材料与制品鉴定"、"质量控制"]

技术转让 Jìshù zhuǎnràng

将一个领域、经济部门或国家的技术有效地和恰当地应用到另一领域、经济部门或国家中去。

为租户的服务 Wèizūhùde fúwù

由房主或其雇员为租户或租借人提供的服务,如维修工作等。

经济公寓 Jīngjì gōngyù

在美国,指包括多个居住单元的建筑物。此词原指美国老城市内早期的公寓住宅,现在是贬义词,多指衰落地区的破旧公寓。[参见"公寓"、"公寓楼"、"危房"、"贫民窟"]

万元定额 Wànyuán dìng'é

中国材料定额的一种,指每万元建筑安装工作量

specifies the amount of material necessary for every ten-thousand-yuan unit of construction work. This norm is used before the preliminary design has been prepared to estimate approximately the amount of material needed for a construction project of a certain scope, as reflected by the planned investment. See also NORM and NORM FOR ESTIMATING MATERIAL REQUIREMENTS.

Tenure

The act of using, as well as the terms and duration for use or possession of land or property.

Territorial planning

In China, the highest level of planning (at a level above regional planning) for the exploitation, utilization, management, and conservation of all natural, labor, and economic resources whether on, in, or under land, sea, or the continental shelf under the territorial sovereignty of China. See also CONSERVATION and RESOURCE MANAGEMENT.

Threshold analysis

Study of the limitations or barriers to expansion that occur at some stage in the growth of a city. According to the theory of threshold analysis

所必须消耗的材料数量，在建设项目初步设计出来之前，根据工程的投资额粗略估算材料用量。[参见"定额"、"材料定额"]

占用权、占用期 Zhànyòngquán、zhànyòngqī

指使用或占有房地产的行为或限期。

国土规划 Guótǔ guīhuà

在中国，为了开发、利用、管理和保护中国领土以内地上、地下、海洋或大陆架的自然和人力经济资源而编制的最高一级（在区域规划之上）的规划。[参见"保护"、"资源管理"]

临界分析（门槛理论） Línjiè fēnxī（Ménkǎn lǐlùn）

对于在城市成长过程中某些限制其发展的极限或障碍的研究。根据由波兰人B.马利兹所提出的临界分析理论，这种极限称为发展的临界。可分为

originally developed by B. Malisz in Poland, the limitations are called thresholds of development. They can be classified as physical, i.e., those created by natural features; technological, i.e., those related to the infrastructure system; and structural, i.e., those requiring reconstruction of parts of the city such as the city center. Through comparison of costs and respective population numbers for all relevant alternatives of possible city expansion, threshold analysis seeks to identify the least costly alternative for overcoming a series of successive thresholds.

Time-sharing arrangement

1. In the United States, the practice of subdividing the right to use real property, especially vacation housing, among a number of individuals. Payment of a percentage of the property's value entitles each person to use the property during a particular time period.

2. Rental of computer time on the same computer by a number of organizations, all of which pay for time used and gain access through telecommunications equipment.

Title

1. Legal documents proving ownership of land or property by a particular individual or organization.

实体上的，指由自然环境造成的；技术上的，指与基础设施系统有关的；结构上的，指城市中某些部分（如市中心）需要重建。该方法力图通过对造价的比较和对城市扩充的各种可能途径所造成的人口数量的比较，找出最经济的途径来克服一系列极限因素。

"定时分享"安排 Dìngshí fēnxiǎng ānpái

1. 在美国，指几个人分享房地产的使用权。每人按该房地产价值的一定百分比投资，从而获得在一定时期内使用该房地产的权利。此种方法多用于度假住房。

2. 指由几个机构或组织共同租用电脑的安排。按使用时间多少支付费用，并通过电讯设备与电脑取得联系。

1. 产权证 Chǎnquán zhèng

在美国，指证明某一个人或组织对土地或产业拥

See also ASSIGNMENT OF TITLE.

2. A section of a particular law, often cited to specify programs provided for under that law, e.g., Title I-Community Development Programs, which are authorized under the U.S. Housing and Community Development Act of 1974.

Top management

See MANAGEMENT STRUCTURE.

Topographic survey

Measurement and mapping of the relief and physical features of a region to show their relative positions and elevations.

Tourist capacity

The maximum number of tourists which a scenic district or spot can accommodate at the same time.

Town

1. In China, a population center smaller than a city but larger than a village. Such settlements were formed historically in connection with markets, transportation facilities, or gathering and distribution of commodities. Administratively, a town may be a subdivision of a county or be at the county level itself, with or without its own government.

有所有权的法律性文件。[参见"过户"]

2. 条款 Tiáokuǎn

在美国,又指一条法律的一部分,用以列出该法律各章节的名称。如1974年美国《住房和社区发展法》中的第一条——"社区发展计划",即为条款。

上层管理 Shàngcéng guǎnlǐ

[见"管理结构"]

地形测量 Dìxíng cèliáng

将某一区域的地形和实物特征加以测量并绘制成地图,以表示其相对位置和标高。

游客容量 Yóukè róngliàng

风景区或风景点所能同时容纳游客的最大数量。

镇 Zhèn

1. 在中国,指小于城市而大于自然村的基层行政区或单位,是由历史或集市、交通、物资集散等原因形成的居民集中点。分县辖镇和县级镇,可设或不设镇政府。

2. In the United States, an urban settlement smaller than a city. In certain regions, a town is a specific type of governmental unit.

Townhouse

In the United States, one of a series of two-to-four-story living units which share common exterior side walls, are designed with a unified architectural plan, and have a separate entrance for each unit. Once built only in cities and towns, townhouses are today found in many suburban and rural areas, providing a moderately priced, space-efficient housing alternative. See also COMMON-AREA MAINTENANCE and ROWHOUSE.

Township

1. In China, a grass roots administrative division under a county, with its own government. Until the recent separation of government administration and commune management, township government was part of the sphere of responsibility of the people's commune.

2. In the U.S. public lands survey, a territorial subdivision six miles long and six miles wide containing thirty-six sections, each one square mile in area. The term township may also refer to a territorial administrative unit which varies widely in its size and

2. 在美国，指小于城市的聚居地。在某些地区，亦指地方政府的一种。

市镇联立式住宅　Shìzhèn liánlìshì zhùzhái

在美国，指一排两至四层联结在一起的住宅单元中的一个。各单元共用外墙，有统一的平面设计，并均有独用户门。以往此种住宅仅建于市镇，现亦建于郊区和农村地区，是一种造价适中、又充分利用空间的住房。[参见"公用部分的维护"、"联立式住宅"]

乡　Xiāng

1. 在中国，指由县管辖的基层行政区域单位，设乡政府。在政社分开以前，乡政权归入人民公社。

2. 根据美国公共土地测量的定义，指一个长宽各为六英里的地域，其中包括 36 个各为一平方英里的地段。但亦可指地区性的行政单位，其大小和结构在美国各地均不相同。

structure in the different regions of the United States.

Town/township enterprise

In China, a collectively owned production or service enterprise run by a town or township in a rural area. Before the establishment of townships and the separation of government administration and commune management, such an enterprise was called a commune/production brigade enterprise.

Traffic forecast

Prediction of the anticipated volume of traffic for various vehicles at a given location and time based on previous vehicle counts at the same location. See also NETWORK ANALYSIS.

Traffic management

The supervision of all transportation functions required by a city or organization in the course of its activities.

Traffic zone

In the United States, a section of a street, road, or highway marked according to certain traffic regulations, e.g., speed limit or prohibition of passing.

乡镇企业 Xiāngzhèn qǐyè

在中国农村地区，由乡镇兴办的生产或服务性企业，一般为集体所有。在政社分开、恢复乡建制前，称为社队企业。

交通预测 Jiāotōng yùcè

根据某一地点以往通过车辆的数量，预测今后该地点在一定时间内各类车辆的交通量。[参见"网络分析"]

交通管理 Jiāotōng guǎnlǐ

对在一个城市或机构的活动过程中所需的各种运输手段进行的监督。

交通管制段 Jiāotōng guǎnzhìduàn

在美国，指根据某些交通规定受管制的街道或公路的一段，例如限制行驶速度、禁止通行等。

Transfer of domicile registration

In China, the change of rural to urban domicile registration or vice versa. This often occurs in suburban areas of cities when a tract of agricultural land is to be requisitioned. With the approval of the city government, a small number of rural inhabitants engaged in farming may be granted permission to be registered in a city and offered jobs in urban areas. The term does not apply to urban residents who move from one city to another as conditions require. See also DOMICILE REGISTER, LAND ACQUISITION, REGISTERED RESIDENT, and RESIDENT REGISTRATION.

Transferable development rights (TDR)

A land management device based on the principle that the development potential of privately held land, i.e., "development rights," is in part a community asset. Government may allocate the asset to enhance the general welfare. In concept, TDR severs the development potential from the land and treats it as a marketable item. A planning authority using TDR establishes conservation zones and transfer zones. The transfer zones are areas deemed suitable for intense development, based on planning theory, available public facilities and utilities, and overall compatibility with both the built and the natural environment. In

转户 Zhuǎnhù

在中国,指由城市户口转为农村户口,或由农村户口转为城市户口。在城市郊区征用农业用地时,经市政府批准后,在该片土地上从事农业生产的一部分农业居民可转为城市户口,并在城市中分配工作。转户不包括城市居民从一城市迁徙到另一城市的户口迁移。[参见"户籍"、"征地"、"户口"、"户口登记"]

可转让开发权 Kězhuǎnràng kāifāquán

根据私有土地的开发潜力(即开发权)是一项社区资产这一原则制定的土地管理方法。政府可划拨此项资产以提高公众福利。从概念上讲,可转让开发权是将开发权和土地分离,成为一项可以买卖的资产。规划当局运用这一方法确定保留现状区和建设区。根据城市规划理论,建设区具有公共设施和市政公用设施条件,以及与已有建设和自然环境的协调等因素,适合于密集型开发,而保留现状区则不允许开发。因此,保留现状区内的开发权与土地脱离,建设区则成了接受开发权的地区,从而密度也可以转移。建设区的最大密度可通过购买保留现状内地产主的开发权而增

contrast, development is not allowed in conservation zones. The development potential is therefore detached from the land parcels in the conservation zones, and transfer zones become the receiving areas for that development potential – thus, transferable density. The maximum density for transfer areas may be exceeded through purchase of development rights from conservation zone landowners, who are thus compensated for the restrictions on development of their land. See also DENSITY TRANSFER and LAND USE CONTROLS.

Transition zone

In the United States, an urban area or zoning district that separates zones of different intensity. The zone may be planned, e.g., a district between a business and a residential district that is zoned for both residential and limited commercial use, or unplanned, e.g., a residential area adjacent to the central business district that has gradually been invaded by business and light industry. See also ZONING DISTRICT.

Transitional neighborhood

In the United States, an urban residential area undergoing a significant change in the composition of its residents and the condition of property. Such neighborhoods are frequently located between a

大。开发受到限制的地产主也可从中得到补偿。
[参见"密度调剂"、"土地使用控制"]

过渡区 Guòdùqū

在美国,指处于不同密集程度地区之间的城市地区或区划地区。过渡可以是经过规划的,如位于住宅区和商业区之间,经规划用于居住和有限商业活动的区;或未经规划的,如邻近商业中心、逐渐被商业和轻工业所侵入的居住区。[参见"区划市区"]

演变中的邻里 Yǎnbiànzhōngde línlǐ

在美国,指居民构成和地产状况处于重大演变中的都市居住区。一般位于维护较好而又稳定的邻里和贫民窟之间,可能向二者任何一种情况演变。[参见"衰退邻里"]

reasonably well-maintained, stable neighborhood and a slum area, and their character may shift in either direction. See also DECLINING NEIGHBORHOOD.

Triplex house

Residential building containing living units for three families, each with its own separate entrance. See also ATTACHED HOUSE.

Turnkey project

In the United States, construction and delivery of a complete facility by a contractor under contract to a government agency or an approved private agency. With little or no involvement by the purchasing agency, the contractor provides financing for the design and construction phases and retains ownership of the project site until construction is complete. Upon completion, the keys are turned over to the owner – hence the name – and full payment is made. See also DESIGN-BUILD PROCESS and PACKAGE DEAL.

三联式住宅 Sānliánshì zhùzhái

指具有三个居住单元供三家使用的居住建筑,各单元有独用户门。[参见"毗连住宅"]

交钥匙工程 Jiāoyàoshi gōngchéng

在美国,指承包人与政府机构或认可的私人机构签订合同,承担一项设施的全部建造过程直至交付使用。承包人负责提供设计和施工阶段所需资金,并保持对在施工程的所有权直至竣工为止。在此期间,买方很少参与或不参与。工程完工后,承包人将全部钥匙交给业主,业主同时付清全部工程价款,这就是交钥匙工程名称的由来。[参见"设计兼施工"、"一揽子交易"]

U

UDAG

See URBAN DEVELOPMENT ACTION GRANT.

Unallotted household

See HOUSING SHORTAGE.

Unauthorized construction

Construction work or structures not licensed by the urban or appropriate local planning authority, or not in conformance with applicable codes and ordinances. See also CODE ENFORCEMENT.

Underground economy

Unofficial purchase and sale or exchange of goods and services, without regard for official market prices, government regulations, licensing requirements, or tax laws.

Underwriting

Guaranteeing a mortgage or offering a loan, and the analysis leading up to this action. The term derives from the official signature symbolically authorizing the loan or guarantee. A loan application is only underwritten after analysis of the risk for a loan applicant and the matching of that risk to an appropriate rate and term. Underwriting is usually

城市开发活动赠款（缩写） Chéngshì kāifā huódòng zèngkuǎn (suōxiě)

［见"城市开发活动赠款"］

无房户 Wúfánghù

［见"住房短缺"］

违章建筑 Wéizhāng jiànzhù

未经城市有关规划部门批准或不符合规范条例要求的建筑物。［参见"执行法规"］

地下经济 Dìxià jīngjì

不遵守正式的市场价格、政府规定、领取执照手续或税收法律等而进行货物或服务事项的非正式买卖或交换。

风险担保 Fēngxiǎn dānbǎo

指对发放抵押贷款或提供贷款所做的分析。此词原指同意发放贷款时由负责人正式具名担保。在此之前必须对贷款申请人的偿还能力进行风险分析，并据此定出相应的贷款利率及条件。这项工作通常由一些专业人员对申请抵押贷款者的信用、偿还能力及财产的可靠性进行分析。［参见"抵押贷款"、"红线注销"］

performed by professionals who assess whether loan applicants are credit-worthy and able to meet payments as well as whether the property will sufficiently secure the loan. See also MORTGAGE and REDLINING.

Unified housing development

In China, a housing development funded by the city government or organizations in different sectors or departments and executed with unified management of planning, design, land acquisition, and construction by the city government or a housing development corporation.

Uniform Construction Index

In the United States, a published, recognized system for ordering building specifications, filing technical information, and cost accounting in construction contract documents.

Unimproved land

A tract that has no infrastructure and has not been built upon or developed for use in any way.

Unit cost

The cost of producing any product on a per-unit basis. In the construction industry, the unit measure

统建住房 Tǒngjiàn zhùfáng

在中国，指由城市或各系统、各部门投资，并由市政府或住房开发公司统一组织规划、征地、设计、建造的住房建设。

统一建筑指标 Tǒngyī jiànzhù zhǐbiāo

在美国，指一种公开发布和公认的指标体系，适用于建筑规格说明、立档技术资料和施工合同文件中的成本核算。

未改善土地 Wèigǎishàn tǔdì

一片没有基础设施和尚未开发的土地。

单位成本 Dānwèi chéngběn

以标准单位计算的产品成本。在建筑业中，指每平方米或每平方英尺建筑面积的工程成本。[参见

for construction cost is generally the square foot or square meter of floor area. See also COST-PRICE INDEX.

Unit price

The money amount for which a product or service is offered for sale, defined as a price per standard unit. See also COST-PRICE INDEX.

Upgrading

Improvement of quality of a structure or neighborhood. This can be accomplished in several ways, e.g., by strict code enforcement or by new community development efforts emphasizing self-help programs.

Urban design

The process of defining the general special arrangement of the activities and objects of a city, with special attention to its attractiveness and amenities. In the complex process of urban design, emphasis is placed on the interrelationship between the physical elements of an urban area and the socio-psychological well-being of its inhabitants. The objective is to create a physical environment which, through its treatment of space and its images, pleases the senses and inspires a spirit of community. See also CITYSCAPE and ENVIRONMENTAL DESIGN.

"成本价格指数"]

单价 Dānjià

每一标准单位产品或服务项目的出售金额。[参见"成本价格指数"]

提高质量 Tígāo zhìliàng

指改进建筑物或邻里的质量。可用几种方法进行,如严格执行法规、新社区的开发中着重推行自助建房计划等。

城市设计 Chéngshì shèjì

指确定一个城市的活动与目标的总体空间布局,使其具有吸引力并使人感到赏心悦目。城市设计的重点在于实体部件的安排有益于居民的社会心理健康。通过对空间及其形象的处理,创造一种物质环境,使居民感到愉快和融洽。[参见"城市风貌"、"环境设计"]

Urban Development Action Grant (UDAG)

In the U.S., the primary federal program to encourage innovative financing for urban redevelopment activities. Awarded through a national competition between local governments, UDAG grants cover a variety of commercial, industrial, or residential projects. A local government must demonstrate both the need for the UDAG and major private financial commitments in a particular project, with a minimum ratio of 2.5 private dollars to each UDAG dollar. This essential "leveraging" principle requires a close partnership between public officials and the private business community. Since the program was begun in 1978, about $4 billion in UDAG grants to over 1, 100 jurisdictions have attracted $23.5 billion in private investment. See also GRANT, LEVERAGING, and URBAN REDEVELOPMENT.

Urban ecology

A theoretical framework which views the urban community as an urban ecosystem analogous to the natural ecosystem. The urban ecosystem, however, is more complex because it encompasses both the interactions between people and of people with their environment, natural and built. Despite theoretical and practical difficulties of reconciling ecological equilibrium with dynamic urbanization, the concept

城市开发活动赠款 Chéngshì kāifā huódòng zèngkuǎn

在美国,指为城市再开发活动而鼓励地方主动集资的联邦计划。各地方政府通过全国性竞争得到财政资助。该项赠款可用于各种商业、工业或居住项目。地方政府须说明对该款项的需要和对某一建设项目的私人投资的规模。一般为每美元赠款需有不少于2.5美元的私人投资。这种杠杆调剂的规定是要求政府官员与私人企业社团之间密切合作。从1978年开始执行这一计划以来,已有约40亿美元赠款发放给1100多个城市管辖地区,并吸引了235亿美元的私人投资。[参见"赠款"、"杠杆作用"、"城市改造"]

城市生态学 Chéngshì shēngtàixué

一种把城市社会比作与自然生态系统相似的城市生态系统的理论。但城市生态系统更为复杂,因为它既包括人与人之间,也包括人与自然和建设环境之间的相互作用。更把动态的城市化和生态平衡协调一致,在理论和实际上都有困难,但城市生态系统这一概念对城市建设的分析是有用的。把人、植物、动物与环境置于一个统一体中,可观察分析其间的相互作用。这种综合的途径可以

of an urban ecosystem is useful in the analysis of urban development. By placing man, plants, animals, and the environment in a single framework within which interactions between elements can be observed and analyzed, the comprehensive approach provides a common language for a number of disciplines. The rational and functional structure of the ecosystem permits identification and quantification of interaction processes within a general system, and consequently, empirical testing of the system by the urban model builder. See also ECOLOGY, DESIGN WITH NATURE, and HUMAN ECOLOGY.

Urban economics

The branch of economics that analyzes the role of the city as an economic unit, serving as a communication center and marketplace for a collection of businesses and industries. As an extension of Location Theory, urban economics focuses on such issues as the factors involved in choice of residence and in location of businesses as well as development of urban transportation networks. See also LOCATION THEORY.

Urban form

The overall physical organization or environment and the special distribution of activities of a city.

为多种学科提供共同的语言。生态系统的合理有效结构可以对整个社会结构中相互作用的过程起鉴别和定量的作用，从而为城市模式的建造者提供检验这个社会结构的机会。[参见"生态学"、"适应自然的设计"、"人类生态学"]

城市经济学　Chéngshì jīngjìxué

经济学的一个分支，把城市作为一个工商业交流中心和市场的经济单位来分析它的作用。城市经济学是"位置理论"的延伸，其研究重点是住处的选择、商业区的位置、城市交通网的发展等问题所涉及的各种因素。[参见"位置理论"]

城市形态　Chéngshì xíngtài

指一个城市的全面实体组成，或实体环境以及各类活动的空间分布。

Urban fringe

The residential area which develops without plan at the edges of a city. The term may be applied neutrally to an outer urban area in which agricultural and residential land uses are mixed but may also refer to outer areas which include poorly constructed, substandard housing without adequate public facilities.

Urban homesteading

See HOMESTEADING.

Urban inhabitant

In China, a person who resides in a city and is officially registered as a resident there. See also REGISTERED RESIDENT, RESIDENCE CERTIFICATE, and RESIDENT REGISTRATION.

Urban network

All towns in a given geographical area with multiple connections among each other and the surrounding rural zones, including interrelationships with other urban areas.

Urban planning

A detailed method or process that seeks to guide the design and development of the built environment,

城市边缘　Chéngshì biānyuán

指在城市外围未经规划形成的居住区。可以是指城市远郊农业和居住用地混杂地带的无褒贬的中性词，也可指住房质量差、不合标准并缺乏足够的公共设施的城市外围地区。

城市定居　Chéngshì dìngjū

［见"公有房地产转让"］

城市居民　Chéngshì jūmín

在中国，指在城市居住并具有城市户口的居民。［参见"户口"、"户口簿"、"户口登记"］

城市网架　Chéngshì wǎngjià

在某一地理区域内所有城市相互之间和他们与周围农村地区之间的多方面联系，包括与区域以外其他城市地区的关系。

城市规划　Chéngshì guīhuà

指预测城市的发展并管理各项资源以适应其发展的具体方法或过程。传统的城市规划多注意城市

anticipating growth and managing resources to accommodate that growth. Traditional planning concentrated on the physical characteristics of the urban region. Modern planning, however, attempts to understand the effects of diverse economic, social, and environmental factors on changing land use patterns and to develop plans that reflect this continual interaction. In the United States, comprehensive planning and preparation of a master plan are key elements in urban planning. In China, urban planning usually consists of comprehensive planning and detailed planning. In some large and medium-sized cities, an intermediate category, city district planning, is added between the two. See also CITY DESTRICT PLANNING, COMPREHENSIVE PLANNING, and DETAILED PLANNING.

Urban policy

A direction or course of action set by some level of government to guide present and future decisions concerning urban development in all its major aspects – economic, social, environmental, physical, and management.

Urban redevelopment

The general improvement, particularly in physical terms, of an older urban area by a variety of means:

地区的实体特征。现代城市规划则研究各种经济、社会和环境因素对土地使用模式的变化所产生的影响，并制订能反映这种连续相互作用的规划。在美国，总体规划总图是城市规划中的关键组成部分。在中国，城市规划通常包括总体规划和详细规划两个阶段。在某些大中城市，在两阶段之间增加城市分区规划。[参见"城市分区规划"、"总体规划"、"详细规划"]

城市政策 Chéngshì zhèngcè

为了从经济、社会、环境、实体及管理等主要方面，指导当前和将来的城市发展决策，而由某一级政府机构制订的方向或一系列行动。

城市改造 Chéngshì gǎizào

利用来自公、私财源的资金，以不同的方法，对旧城进行改造，尤其是在实体方面，包括建造新

new construction; building rehabilitation, including adaptive re-use; neighborhood conservation; historic preservation; and the upgrading of infrastructure. Funding comes from a variety of public and private sources. Today this term is closely linked to the economic and social rejuvenation of a city and is often used interchangeably with urban revitalization. See also ADAPTIVE RE-USE, HISTORIC PRESERVATION, NEIGHBORHOOD CONSERVATION, REDEVELOPMENT, REHABILITATION, URBAN DEVELOPMENT ACTION GRANT, URBAN RENEWAL, and URBAN REVITALIZATION.

Urban renewal

Process through which deteriorated neighborhoods are upgraded through clearance and redevelopment (buildings, infrastructure, and public amenities). Building code enforcement and property rehabilitation are also prominent in renewal areas. Urban renewal activities can be financed with a combination of federal and local funds or strictly with private funds. An active federal urban renewal program (1949-1974), through its grants and loans, had a great effect on the redevelopment of several U.S. cities. However, as a result of widespread criticism of its disruptive effects on the social life of

的建筑物、将旧建筑修复再利用或改作他用、邻里保护、历史性保护及改进基础设施等。本词含义与恢复一个城市的经济和社会活力相近,并与"城市复苏"通用。[参见"适应性再使用"、"历史性保护"、"邻里保护"、"改造"、"复兴"、"城市开发活动赠款"、"城市更新"、"城市复苏"]

城市更新 Chéngshì gēngxīn

指通过清除和改造房屋、基础设施和宜人事物而对衰退邻里进行改造。更新的地段内的一个特点是贯彻执行建筑法规和对房产修复后重新利用。城市更新活动可由联邦和地方财政共同投资,或仅由私人投资,1949—1974年的联邦城市更新计划,以其赠款与借款,给美国若干城市的改造带来卓著效果。但是,由于破坏了邻里社会生活而招致广泛的批评,使得该词现在不再普遍使用,而代之以"城市改造"。[参见"城市改造"]

neighborhoods, the term "urban renewal" is not now commonly used, having been largely replaced by "urban redevelopment." See also URBAN REDEVELOPMENT.

Urban revitalization

A term very close in meaning to urban redevelopment, but especially emphasizing urban economic development strategies such as public-private partnerships to create new jobs and to restore a positive public attitude toward a city's future. Examples of successful urban revitalization efforts are the programs undertaken in Baltimore, Maryland, and Boston, Massachusetts, in 1960 to 1985. See also HOMESTEADING, NEIGHBORHOOD REVITALIZATION, and URBAN REDEVELOPMENT.

Urban sociology

The study of the development, structure, and interaction of human groups and institutions within an urban environment.

Urban sprawl

Spatial expansion of an urban settlement in a disorganized, haphazard manner without advance planning or regard for transportation and service needs. See also AGGLOMERATION.

城市复苏 Chéngshì fùsū

与城市改造含义相近,但侧重于城市经济发展战略。如通过公、私双方合作以创造新的就业机会,并使公众对城市日后的发展恢复积极态度。巴尔的摩(马里兰州)、波士顿(马萨诸塞州)1960—1985年的复苏计划均为较成功的例子。[参见"定居耕种"、"邻里复苏"、"城市改造"]

城市社会学 Chéngshì shèhuìxué

研究城市环境内人类集团和机构的发展、结构和相互作用的科学。

城市无计划扩展 Chéngshì wújìhuà kuòzhǎn

指城市空间在无组织、无事先计划、无视交通和服务设施等需要的情况下的盲目扩张。[参见"聚集体"]

Usable floor area

In China, the net floor area in a dwelling unit for the use of the residents, i.e., living floor area plus service floor area. See also LIVING FLOOR AREA and RESIDENTIAL FLOOR AREA.

User fee

In the United Stated, charge to an individual or organization for any services, but especially for water supply, sewage, and garbage collection. Such fees are often imposed by government as a means of assessing citizens' demand for particular services, tempering overuse of certain services, and generating additional revenues. Also called a benefited user charge. See also CAPACITY BUILDING and FEE.

User needs

See USER SURVEY.

User survey

The collection and analysis of data on individuals' experiences with and reactions to systems, services, or products that they are currently using. A user survey may also be designed to determine individuals' interest in or need for systems, services, or products under development.

使用面积　Shǐyòng miànjī

在中国，指居住单元内供居住者使用的建筑净面积，即居住面积加辅助面积。[参见"居住面积"、"居住建筑面积"]

用户费　Yònghùfèi

在美国，指私人或机构为取得某项服务所付的费用，尤指供排水、收集垃圾等费用。通常由政府规定，是用以测定市民对某种服务的需求、限制过度使用并增加税收的一种方法。又称"受益费"。[参见"扩大能力"、"费"]

用户需求　Yònghù xūqiú

[见"用户调查"]

用户调查　Yònghù diàochá

收集分析用户对现时使用的某些系统、服务或产品的意见的数据。用以了解用户对某些系统、服务以及正在生产的产品的兴趣和要求。

Utilities

See INFRASTRUCTURE.

Utilities engineering

Coordinated disposition of various underground lines or pipe systems for public utilities, taking into consideration directions of roads and lines, depth of lines below ground level and variations in level, distances between lines, different types of pits and holes, common trenching, and so forth.

Utility costs

The charges for basic services to a dwelling, including water, electricity, natural gas or other heating and cooking fuels, garbage collection, and sewage disposal, but excluding telephone service. In the United States, utility costs are an important consideration in estimating the overall cost of renting or purchasing a dwelling.

Utility survey

A listing of the existing utilities on a construction site.

公用设施 Gōngyòng shèshī

［见"基础设施"］

管网综合 Guǎnwǎng zōnghé

城市市政设施中各种地下管网的综合布置，包括道路与管道走向、地下埋深与标高变化、管道间距、各类井、孔位置及通用管沟等。

公用设施费用 Gōngyòng shèshī fèiyòng

住房中基本服务项目的费用，包括水电费、天然气或其他采暖和厨房燃料费、垃圾收集费等，但不包括电话费。在美国，公用设施费用是在估算租用或购买住房的全部费用时一个重要的考虑因素。

公用设施调查 Gōngyòng shèshī diàochá

对建筑工地上现有公用设施开列清单。

V

Vacancy rate

The proportion of unoccupied houses or apartment units compared to the total number of existing units. See also HOUSING SHORTAGE and OCCUPANCY RATE.

Value engineering

Systematic use of techniques to identify the required function of a product or system, to establish a value for the function, and to provide the function at the lowest overall cost. In the housing industry, value engineering is a total systems approach to selecting house designs, materials, and products that represent the least costly marketable combination. Also called value analysis and value control.

Van pool

See CAR POOL.

Variable rate mortgage

See ADJUSTABLE RATE MORTGAGE.

"Vernacular housing"

In China, dwellings in a traditional local style peculiar to a particular region. Vernacular houses may be located in either rural or urban areas and are usually built or inherited by the owners. See also

空房率 Kōngfáng lǜ

指无人居住的住宅或公寓在住房单元总数中所占的百分比。[参见"住房短缺"、"住房占用率"]

价值工程 Jiàzhí gōngchéng

通过系统地采用各种技术，鉴定某种产品或体系的功能，从而确定该功能的价值，并以最低价格提供该种功能。在住房业中，价值工程是用来对住房设计、材料和制品进行综合考虑并做出符合市场最低价格的选择。又称"价值分析"、"价值控制"。

合伙乘车 Héhuǒ chéngchē

[见"合乘"]

可变利率抵押贷款 Kěbiàn lìlǜ dǐyā dàikuǎn

[见"可调利率抵押贷款"]

民居 Mínjū

在中国，指具有地方传统风格的居住建筑。一般为城镇或农村中由私人自建或前人留下的住宅。[参见"四合院"]

COURTYARD HOUSE.

Vested right

An absolute and unconditional right of an individual that is usually protected by a constitutional guarantee and cannot be abrogated by action of any other private individual.

"Vest-pocket park"

In the United States, a small scenic or recreational area built in available space between buildings or on a vacant lot in a city.

Village

In China, a rural settlement, usually with 500 to 800 inhabitants, that has evolved historically rather than through planned development.

Voluntary security service

In China, a voluntary assistance organization of urban inhabitants for protecting public security. The service supports the police substation and neighborhood committee by reminding inhabitants to take security precautions or by patrolling. For a U.S. organization with somewhat analogous functions, see also NEIGHBORHOOD WATCH.

固有权利 Gùyǒu quánlì

一种个人所拥有的绝对和无条件的权利,受宪法保障,不得被任何他人的行动所取消。

"口袋公园"(小游园) Kǒudài gōngyuán (Xiǎoyóuyuán)

在美国,指在建筑物之间或城市空地上所建造的小型风景或游乐地区。

自然村 Zìráncūn

在中国,指农民因生产原因世代聚居而形成的村落,区别于通过规划建设形成的定居点,一般拥有人口500—800人。

居民联防 Jūmín liánfáng

在中国,指城市居民配合派出所和居委会进行巡逻或提醒住户注意治安的自愿组织。[美国类似性质的组织可参见"邻里自卫"]

W

Waiver of regulations

Exemption from or modification of regulatory requirements for a particular building project. Also referred to informally as relaxation of regulations.

Waste management

The planning, coordination, and supervision of the disposal of refuse and wastewater from human populations and industry. See also ADVANCED WASTEWATER TREATMENT, PRIMARY WASTEWATER TREATMENT, SECONDARY WASTEWATER TREATMENT, SEWAGE DISPOSAL SYSTEM, and WASTE UTILIZATION.

Waste utilization

Treatment and use of residential sewage, usually in the form of sludge, for landfill, fertilizer, or other purposes. See also ADVANCED WASTEWATER TREATMENT, PRIMARY WASTEWATER TREATMENT, SECONDARY WASTEWATER TREATMENT, SEWAGE DISPOSAL SYSTEM, and WASTE MANAGEMENT.

Wetlands

In the United States, a land area partially or completely covered by water, swamps, and bogs, and frequently protected by law from the adverse effects

放宽条例 Fàngkuān tiáolì

允许在某项目工程上变通或免予遵守某些规定,亦可称"放松条例"。

废物管理 Fèiwù guǎnlǐ

对人类和工业所排放的废物和废水处理所做的规划、协调和监督工作。[参见"污水深度处理"、"污水一级处理"、"污水二级处理"、"污水处理系统"、"废物利用"]

废物利用 Fèiwù lìyòng

指采用污泥法对民用污水的处理和使用。污泥可用于填土和施肥等。[参见"污水深度处理"、"污水一级处理"、"污水二级处理"、"污水处理系统"、"废物管理"]

塘地 Tángdì

在美国,指部分或全部被水、沼泽或泥塘覆盖的土地。由于具有保护野生动物、植物以及控制水泛和游乐的价值,塘地多受法律保护,以防开发

of development because of its value for wildlife preservation, flood control, and recreation.

Work element construction program

In China, a directive document with monthly and ten-day work plans for the basic level of the construction enterprise, i.e., the construction brigade. Prepared on the basis of the construction management plan and actual onsite conditions, the work element construction program is used to assign work to appropriate specific individuals in annual and quarterly plans, thus ensuring rational distribution and full utilization of labor and material resources.

Working drawing

Any plan showing sufficient detail to permit construction of the object or project without other drawings or instructions. See also CONSTRUCTION DOCUMENTS.

Working drawing estimate

See DETAILED ESTIMATE.

"Write-down costs"

In the context of U.S. urban redevelopment, the difference between the cost of purchasing and clearing redevelopment land, and the price at which

带来不良后果。

施工作业计划 Shīgōng zuòyè jìhuà

在中国,指指导建筑安装企业的基层施工单位(工程队)完成施工任务的文件,包括月作业计划和旬作业计划。施工单位参照施工组织设计并根据现场的实际情况,将年度、季度的计划任务具体分配到有关的执行者,使人力、物力得到合理调配和充分利用,以保证施工任务按计划完成。

施工图 Shīgōngtú

一种细节交代清楚、所示全部内容不需其他图纸或说明书即可按图施工的图纸。[参见"施工文件"]

施工图预算 Shīgōngtú yùsuàn

[见"设计预算"]

"减记成本" Jiǎnjì chéngběn

按美国城市改造的含义,是指购买和清理需要改造的土地的费用和把清理过的土地卖给开发者的价格之间的差额。由于该种土地是以低于市价的

the cleared land is sold to developers. As the land is sold at below-market prices, the financial loss is assumed by the land development authority. See also REDEVELOPMENT.

价格出售的,财务上的损失由土地开发当局承担。
[参见"改造"]

Y

"Yijing"

One of the principles of traditional Chinese art used in garden design and urban planning. "Yinjing" refers to an artistic atmosphere created by treating the various physical elements of a limited landscape or urban environment as a harmonious composition intended to stimulate one's diverse associations. Thus, a relatively small physical space is made to serve the larger world of Chinese philosophy. See also "BORROWED VIEW," LANDSCAPE ARCHITECTURE, and "SCENIC FOCAL POINT."

意境 Yìjìng

在园林设计和城市规划中运用的中国传统艺术原则之一,即把风景或城市环境中的实体作为一个和谐的整体来处理,旨在创造一种能引起人们联想的艺术境界,从而使一个相对狭小的实体环境体现出中国广阔的哲学世界。[参见"借景"、"风景建筑学"、"对景"]

Z

Zero lot line

Placement of a building on a lot with one wall of the structure on a side property line, allowing for utilization of the remainder of the lot as outdoor living space. In the United States, this design approach is commonly used for high-density residential developments to permit reduced lot size and land costs.

Zoning

In the United States, adoption of municipal ordinances that specify permissible land uses for designated urban districts, as well as exceptions to those uses and procedures for amendment and appeal. The purpose of zoning is to control and direct the use and development of property. Regulations may limit the use of land and buildings, the height and bulk of buildings, the proportion of a lot which buildings may cover, and the density of population of a given area. See also ACCESSORY USE, COMPREHENSIVE PLANNING, DOWNZONING, LAND USE CONTROLS, MIXED-USE ZONING, PERMITTED USE, REZONING, SPOT ZONING, and ZONING DISTRICT.

Zoning administrator

In the United States, a special official with training

地界零线 Dìjiè língxiàn

将房屋的一面外墙置于地界的边线上,从而将余下的地作为户外活动空间。在美国,这种设计多用于高密度住宅区,以压缩用地面积和地价。

区划 Qūhuà

在美国,指一种城市法令。其中除因修正和申诉而允许例外用途外,对每个市区的土地都规定具体的允许用途。其目的在于控制及引导房地产的使用和开发,规定中可对土地和建筑物的使用、建筑物的高度和体积、建筑物的占地面积比和某一地区的人口密度等做出限制。[参见"附带用途"、"总体规划"、"降低密度区划"、"土地使用控制"、"多用途区划"、"许可使用"、"重新区划"、"个别点区划"、"区划市区"]

区划行政官 Qūhuà xíngzhèngguān

在美国,指受过正式城市规划训练、专门负责监

in city planning who supervises enforcement of zoning regulations and must approve applications for building permits and certify compliance before a permit can be issued by a building inspector. See also BUILDING INSPECTOR, BUILDING PERMIT, and CODE ENFORCEMENT.

Zoning amendment

In the United States, changes to zoning regulations or district boundaries, usually by the city council. The power to amend boundaries is frequently limited by state enabling legislation to prevent abuse. See also DOWNZONING, REZONING, and ZONING.

Zoning by right

In the United States, zoning that grants owners the right to use buildings and land for purposes defined in ordinances, for example, accessory use, without special permission of the local zoning authority.

Zoning district

In the United States, a section or zone of a jurisdiction within which permissible uses are prescribed. Restrictions on building height, bulk, layout, and other features are defined by ordinances. See also DISTRICT, SPECIAL DISTRICT, TRANSITION ZONE, and ZONING.

督区划法规实施的官员。在建筑检查员签发施工执照前,其建筑许可证的申请必须由区划行政官批准,并证明该项建筑符合区划规定。[参见"建筑监察员"、"施工执照"、"执行法规"]

区划修正 Qūhuà xiūzhèng

在美国,指区划法规或市区界限的变更,通常由市议会提出。为了便于立法,防止权力滥用,更改界限的权力一般由州政府加以限制。[参见"降低密度区划"、"重新区划"、"区划"]

区划授权 Qūhuà shòuquán

在美国,指房地产主有权根据区划法规中规定的用途使用建筑物和土地,而无须向地方区划当局再申请批准,例如"附带用途"。

区划市区 Qūhuà shìqū

在美国,指一个行政建制范围内具有特定用途的地段或区。在区划法令中对建筑物的高度、体积、布局及其他特点的限制均有规定。[参见"区"、"特区"、"过渡区"、"区划"]

Zoning permit

In the United States, written authorization for an owner to use land or buildings within a zoned district for purposes allowed under existing ordinances.

Zoning variance

In the United States, a special permit that allows a property owner to improve his/her land in a manner not in conformance with local ordinances. See also BOARD OF ZONING APPEALS and SPECIAL USE PERMIT.

区划许可证 Qūhuà xǔkězhèng

在美国，指在区划市区内，允许业主按现行区划法令使用其土地或建筑物的书面凭证。

区划特许证 Qūhuà tèxǔzhèng

在美国，指允许业主不按地方区划法令规定的方式改进其土地的一种特别许可证。[参见"区划申诉委员会"、"专用许可证"]

SUBJECT INDEXES / 主题分类

1. Housing Types and Uses / 住宅类型与用途
2. Real Estate Finance and Residential Mortgages / 房地产资金与住房抵押贷款
3. Construction Management – Process and Personnel / 施工管理——程序与人员
4. Construction Estimates and Costs / 施工预算与成本
5. Bids and Contracts / 招标与合同
6. Area / Density Measures / 面积、密度测度
7. Land Use Planning and Zoning / 土地利用规划与区划
8. Economic and Community Development / 经济与社区建设
9. Urban Development Forms / 城市开发形式
10. Urban Design and Historic Preservation / 城市设计与历史性建筑保护
11. Environment/Infrastructure/Transportation / 环境、基础设施、交通运输

1. Housing Types and Uses / 住宅类型与用途

Accessory housing	合住住房
Apartment house	公寓楼
Assisted housing	公助住房
Attached house	毗连住宅
Boardinghouse	供膳宿的私家住宅
Bungalow	带回廊的小住宅
Cave dwelling	窑洞住宅
Condominium	住户自有公寓
Congregate housing	集合公寓
Core housing	核心住房
Courtyard house	四合院
Detached house	独立式住宅
Duplex apartment	跃层公寓
Dwelling unit	居住单元
Earth-sheltered housing	掩土住房
Elderly housing	老年住房
Emergency housing	应急住房
Entitlement to housing	住房申请权
Family housing	家属宿舍
Garden apartment	花园公寓
Group house	合租住宅
Halfway house	戒瘾教育所
Housing cooperative	住房合作社
Manufactured housing	工厂预制住房
Modular housing	模数制住房
Multifamily housing	多户住房

Municipally owned housing	房管部门住房
Mutual housing	互助住房协会
Obligation of organizations to provide housing	住房分配
Panelized housing	大板式住房
Privately aided public housing	公建民助
Public housing	公共住房
Publicly aided private housing	自建公助
Quadruplex house	四联式住宅
Relocation housing	周转房
Rental housing	租用住房
Replacement housing	拆迁户住房
Rowhouse	联立式住宅
Self-help housing	自建住房
Split-level house	错台式住宅
Squatter settlement	违章居留地
"Stick-built"	木架房屋
Studio apartment	画室公寓
Tenement building	经济公寓
Townhouse	市镇联立式住宅
Triplex house	三联式住宅
"Vernacular housing"	民居

2. Real Estate Finance and Residential Mortgages / 房地产资金与住房抵押贷款

Accelerated depreciation	加速折旧
Adjustable rate mortgage (ARM)	可调利率抵押贷款（缩写）
Amortization	（1）分期偿还 （2）逐步淘汰
Appraisal	估价
Appreciation	增值
Assets	资产
Assisted housing	公助住房
Balloon mortgage	特大尾数抵押贷款
Blanket mortgage	揽总抵押
Break-even point	收支平衡点
Buydown	折扣售房
Carrying charges	土地持有费
Cash flow	现金流量
Closing	结算
Cofinancing	共筹资金
Coinsurance	共同保险
Compensation for land	土地补偿费
Compensatory replacement of demolished housing	原拆原建
Debt service	债务清偿服务
Default	违约
Delinquent mortgage	拖欠抵押贷款
Depreciation	折旧、贬值

Downpayment	首次付款
Drawdown	分期提款
"Earnest money"	定金
Equity	资产净值
Escrow	寄存担保品
Finance	筹资
Fixed-rate mortgage	固定利率抵押贷款
Foreclosure	取消赎回权
Graduated payment mortgage	递增偿还抵押贷款
"Holdback"	暂扣款
Housing allowance	住房津贴
Housing commercialization	住房商品化
Land costs	地价
Lease	租约
Leasehold	租赁
Lease-purchase agreement	租赁购置协议
Liabilities	负债
Loan terms	借款条件
Loan-to-value ratio	贷款与价值比
Market price	市场价格
Market rate loan	市场利率贷款
Market rent	市场租价
Mortgage	抵押贷款
Mortgage bank	抵押贷款银行
Mortgage revenue bond	抵押收益债券

Organization-raised funds	自筹资金
Primary mortgage market	初级抵押贷款市场
Prime interest rate	优惠利率
Privately aided public housing	公建民助
Rate of return	利润率
Refinance	重筹资金
Revolving fund	周转金
Savings and loan association	储蓄贷款协会
Secondary mortgage market	二级抵押贷款市场
Shared appreciation mortgage	分享增值抵押贷款
Sweat equity	改进后净增值
Underwriting	风险担保

3. Construction Management–Process and Personnel / 施工管理——程序与人员

Acceptance of hidden subsurface work	隐蔽工程验收
Acceptance of work subelements	分项工程验收
Advance payment for materials	预付材料款
Architect	建筑师、建筑师事务所
Architect's approval	建筑师认可证明
Building inspector	建筑监察员
Completed item	建成投入生产项目
Completed project	全部竣工项目
Construction brigade	工程队（工段）
Construction company	建筑公司
Construction crew	施工队伍
Construction division	工程处（工区）
Construction enterprise	建筑安装企业
Construction flow process	流水作业法
Construction headquarters	工程指挥部
Construction inspector	施工检查员
Construction management	施工管理
Construction management plan	施工组织设计
Construction manager	施工经理
Construction organization	施工单位
Construction support base	生产基地

Construction team	施工队
Construction work element	分部工程
Construction work quantity	建筑安装工程量
Consultant	顾问
Critical Path Method (CPM)	关键线路法（缩写）
Description of construction progress	工程形象进度
Design-build process	设计兼施工
Detailed construction schedule	施工进度计划
Expediter	器材调度员
Fast track construction	快速施工法
Final acceptance	最终验收
Foreman	工长
Job assignment notice	工程任务单
Job captain	设计组长
Job superintendant	工段长
Nonconforming use	不符规定使用
Overall construction site plan	施工总平面图
Project manager	工程主任
Punch list	建设缺陷清单
Quantity survey	（1）工程用料与设备清单 （2）空地或空房调查表
Resident architect	驻场建筑师
Resident engineer	驻场工程师
Site inspection	现场监督

4. Construction Estimates and Costs/ 施工预算与成本

Area method	面积估价法
Area take-off	粗略估算
Construction budget	工程预算
Construction cost	建筑成本
Construction estimate	施工预算
Construction financing	筹措建造资金
Construction norm	施工定额
Construction payment	建筑造价
Construction work value	建筑安装工作量
Contingency allowance	不可预见费
Cost accounting	成本核算
Cost breakdown	成本分类
Cost controls	成本控制
Cost estimate	成本估计
Cost overrun	成本超支
Detailed estimate	设计预算
Direct expenses	直接费
Extra expenses	独立费
"Gap financing"	资金缺口
"Hard costs"	硬费用
Indirect expenses	间接费
Labor requirements per square meter	单方用工
Material requirements per square meter	单方用料

Norm	定额
Norm for detailed estimates	预算定额
Norm for estimating labor requirements	劳动定额
Norm for estimating material requirements	材料定额
Norm for preliminary estimates	概算定额
Overhead	管理费
Preliminary estimate	设计概算
Project budget	建设项目投资预算
Service fee norm	取费标准
Statutory profit	法定利润
Ten-thousand-yuan norm	万元定额
Unit cost	单位成本
Unit price	单价
Value engineering	价值工程

5. Bids and Contracts / 招标与合同

Accepted bid	中选标
Addendum	补充文件
Alternate bid	备用标价
Application for payment	付款申请书
Approved equal substitution	准许代用品
Base bid	基本标价
Bid	投标
Bid abstract	标价总表
Bid bond	投标保单
Bid price	（1）出价
	（2）投标价格
Bidding documents	招标文件
Bonus and penalty clause	奖罚条款
Certification for payment	付款证明书
Change order	工程变更通知单
Completion bond	完工保证书
Construction budget	工程预算
Contract	合同
Contract administration	合同管理
Contract award	发包
Contract bond	合同担保书
Contract documents	合同文件
Contract item	合同项目
Contractor	承包人
Contractor's option	承包人的选择权
Deposit for bidding	投标文件押金

documents	
Field order	变更通知
Fixed-price competition	固定价格竞争
Fixed-price contract	固定价格合同
General contract	总承包合同
Maintenance bond	维修保证书
Percentage agreement	按百分比收费协议
Performance bond	履行合同保单
Prime contract	直接承包合同
Progress payment	工程进度款
Request for Proposal (RFP)	招标通告(缩写)
Schedule of values	工程分项价值表
Subcontract	分包合同
Turnkey project	交钥匙工程

6. Area/Density Measures/ 面积、密度测度

Average floor area per unit	平均每套建筑面积
Average living floor area per capita	平均每人居住面积
Average number of persons per household	平均每户人口
Buildable area	可建面积
Building area	基底面积
Building density	建筑密度
Building efficiency	可出租面积比
Density of living floor area	居住面积密度
Density of registered inhabitants	居住人口密度
Dwelling size	户型
Floor-area ratio (FAR)	建筑面积比（缩写）
Gross floor area (GFA)	建筑毛面积（缩写）
High-rise/high density	高层高密度
High-rise/low density	高层低密度
Leasable area	可出租面积
Low-rise/high density	低层高密度
Net residential area	居住用地净面积
Net site area	住房基地净面积
Population density	人口密度
Residential density	居住密度
Residential floor area	居住建筑面积
Service floor area	辅助面积

Size distribution of dwellings	户室比
Size distribution of households	户型比
Usable floor area	使用面积

7. Land Use Planning and Zoning / 土地利用规划与区划

Accessory use	附带用途
Aesthetic controls	美观控制
Aesthetic zoning	美观区划
Board of zoning appeals	区划申诉委员会
City district planning	城市分区规划
City planning and administration bureau	城市规划管理局
City planning commission	城市规划委员会
Cluster zoning	组团式区划
Comprehensive planning	总体规划
Council of Governments (COG)	规划协调委员会（缩写）
Density transfer	密度调剂
Designated function of a city	城市性质
Detailed planning	详细规划
Downzoning	降低密度区划
Floating zone	浮动区
Functional planning	功能规划
Height zoning	建筑高度区划
Highest and best use	最佳用途
Incentive zoning	鼓励性区划
Land readjustment	土地调整
Land reclamation	土地整治
Land use assessment	城市用地评价

Land use balance	用地平衡
Land use controls	土地使用控制
Land use plan	土地利用规划图
Land use survey	土地利用调查
Master plan	规划总图
Mixed-use zoning	多用途区划
Nonconforming use	不符规定使用
Optimal land utilization	节约用地
Performance standard zoning	功能标准区划
Permitted use	许可使用
Physical planning	实体规划
Planning commission	规划委员会
Principal use	主要用途
Regional planning commission	区域规划委员会
Rezoning	重新区划
Sectorization of functions	功能分区
Sector planning	局部规划
Segregated use	分区专用
Site development	建设场地开发
Special use permit	专用许可证
Spot zoning	个别点区划
Threshold analysis	临界分析（门槛理论）
Transferable development rights (TDR)	可转让开发权（缩写）
Zoning	区划

Zoning administrator	区划行政官
Zoning amendment	区划修正
Zoning by right	区划授权
Zoning district	区划市区
Zoning permit	区划许可证
Zoning variance	区划特许证

8. Economic and Community Development / 经济与社区建设

Agro-industrial-commercial enterprise	农工商联合企业
Block grant	一揽子赠款
Bond	（1）保证书 （2）债券
Cadastral survey	地籍册
Capital grant	公共工程项目投资
Capital improvement program	资本刷新计划
Capital investment	资本投资
Commercial network	商业网点
Community center	社区中心
Community development corporation (CDC)	社区开发公司（缩写）
Community facility	社区设施
Competitive funding	竞争性拨款
Comprehensive economic equilibrium	综合平衡
Cost-benefit analysis	成本效益分析
Debenture	信用债券
Discretionary funding	选择性拨款
Domestic loan	国内贷款
Economic region	经济区
Enterprise zone	兴业区
Fixed-asset investment	固定资产投资

Foreign loan	国外贷款
Formula funding	按公式拨款
"Gap financing"	资金缺口
Gentrification	中产阶级化
"Hold-harmless" provision	不受损条款
Individual business	个体户
Industrial-agricultural enterprise	工农联合企业
Industrial modernization investment	更新改造措施投资
Industrial modernization project	更新改造措施项目
Investment in capital construction	基本建设投资
Investment tax credit	投资减税额
Joint venture	合资经营
Labor equilibrium	劳动平衡
Leveraging	杠杆作用
Matching grant	相应赠款
Multiplier effect	收入增殖作用
National economic and social development planning	国民经济和社会发展计划
National project	国家项目（部直属项目）
Neighborhood revitalization	邻里复苏
Organization-raised funds	自筹资金
Poverty level	贫困线
Redevelopment	改造

Revolving fund	周转金
Service facility	生活服务设施
Sideline production	副业生产
Sites and services	提供场地与基础设施
Small business	小型企业
Special economic zone	经济特区
State investment	国家投资
Subsidy	补贴
Transferable development rights (TDR)	可转让开发权(缩写)
Upgrading	提高质量
Urban Development Action Grant (UDAG)	城市开发活动赠款(缩写)
Urban redevelopment	城市改造
Urban renewal	城市更新
Urban revitalization	城市复苏

9. Urban Development Forms / 城市开发形式

Agglomeration	聚集体
Bedroom suburb	卧城郊区
Buffer zone	防护带
Central business district (CBD)	商业中心区（缩写）
Central city	中心城市
City proper	城区
Civic center	市政中心
Cluster development	组团式建设
Conurbation	集合城市
District	区
Downtown	闹市区
Exurbia	城市远郊
Freestanding new town	独立新城
Ghetto	少数人种聚居区
Greenbelt	绿带
Growth center	增长中心
Industrial park	工业园
Infill development	填空性建设
Main street	主要街道
Metropolitan area	大都市地区
Mixed-use development	多用途建设
Mixed-use zone	多用途区
Neighborhood	邻里
New town	新城
Piecemeal development	零星建筑

Primate city	初级城市
Residential district	居住区
Shopping mall	购物商场
Special district	特区
Strip development	条形商业区
Suburb	郊区
Transition zone	过渡区
Urban fringe	城市边缘
Urban sprawl	城市无计划扩展

10. Urban Design and Historic Preservation / 城市设计与历史性建筑保护

Adaptive re-use	适应性再使用
Beautification	美化
"Borrowed view"	借景
Certification of historic buildings	历史性建筑保护证明
City Beautiful Movement	城市美化运动
Cityscape	城市风貌
Conservation of Historic Cultural Cities	历史文化名城保护
Conservation of historic landmarks and sites	文物古迹保护
Conservation of scenic spots	风景名胜保护
Culturally sensitive design	文化特色区设计
"Design with Nature"	适应自然的设计
"Edible landscaping"	食用园林布置
Environmental design	环境设计
Greenbelt	绿带
Historic American Buildings Survey	美国历史性建筑调查
Historic District	历史性市区
Historic preservation	历史性建筑保护
Human scale	人类尺度
Landscape architecture	风景建筑学
Neighborhood conservation	邻里保护

Open space	空地
Pedestrian-mall	人行林荫路
Rehabilitation	复兴
Restoration	修复
"Scenic focal point"	对景
Skyline	天际线
Streetscape	街景
Urban design	城市设计
Urban form	城市形态
"Yijing"	意境

11. Environment / Infrastructure / Transportation / 环境、基础设施、交通运输

Advanced wastewater treatment (AWT)	污水深度处理（缩写）
Aquifer	蓄水层
Arterial road	主干道
Artery	干线
Bypass	绕行路
Carrying capacity	持续能力
Coefficient of utilization	利用系数
Collector street	辅助道路
District heating	区域供热
Ecology	生态学
Energy conservation	能源保护
Energy efficiency ratio (EER)	能源效率比（缩写）
Energy performance standard	能量效率标准
Environmental design	环境设计
Environmental impact statement	环境影响报告
Environmental standards	环境标准
Expressway	高速公路
Feeder bus system	公共汽车支线系统
Floodplain	泛滥平原
Freeway	快速干道
Grade separation	立体交叉

Hazardous wastes	有害废物
Heating, Ventilating, Air Conditioning (HVAC)	供暖、通风、空调（缩写）
"Hutong"	胡同
Interstate Highway System	州际公路网
Level crossing	平交
Mass transit	大量客运
Microclimate	小气候
Modal split	交通工具分类
Natural Conservation Zone	自然保护区
Network analysis	网络分析
Noise controls	噪声控制
Noxious industry	有害工业
Origin-destination survey (O and D)	起终点调查
Parkway	林荫公路
Primary wastewater treatment	污水一级处理

SELECTED BIBLIOGRAPHY / 参考书目

English Language References

Abrams, Charles. Language of Cities: A Glossary of Terms. New York: Viking, 1971.

American Institute of Architects. Glossary of Construction Industry Terms. New York: American Institute of Architects, 1982.

Allen, Robert D., and Thomas E. Wolfe. Allen &Wolfe Illustrated Dictionary of Real Estate. New York: John Wiley & Sons, 1983.

Ammer, Christine, and Dean S. Ammer. Dictionary of Business and Economics. New York: The Free Press, and London: Collier Macmillan Publishers, 1977.

Arnold, Alvin L., and Jack Kusnet. The Arnold Encyclopedia of Real Estate. Boston and New York: Warren, Gorham & Lamont, 1978.

Davids, Lewis E. Dictionary of Banking and Finance. Totowa, N. J.: Littlefield, Adams & Co., 1980.

Hanson, John L. A Dictionary of Economics and Commerce. London: MacDonald & Evans Ltd., 1974.

Logie, Gordon. Glossary of Population and Housing: English-French-Italian-Dutch-German-Swedish. Amsterdam: Elsevier Scientific Publishing Company, 1978.

Mortgage Banking Terms: A Working Glossary. Washington, D. C.: Mortgage Bankers Association of America, 1978.

Munn, Glenn G. Encyclopedia of Banking and Finance. 8th ed. Edited by F. L. Garcia. Boston: Bankers Publishing Company, 1983.

Olin, Harold B., and Christina Farnsworth. Housing/Planning

Glossary. Chicago: United States League of Savings Associations, 1979.

Pearce, David W. The Dictionary of Modern Economics. Cambridge, Mass.: MIT Press, 1981.

Roberts, John M. Construction Management: An Effective Approach. Reston: Reston Publish Company, Inc., 1980.

Stein, J. Stewart. Construction Glossary. New York: John Wiley & Sons, 1980.

UIA International Vocabulary of Town Planning and Architecture. Paris: Societe de Diffusion des Techniques du Batimentet des TravauxPubliques, 1970 (In French, English, and German).

Whittick, Arnold. Encyclopedia of Urban Planning. New York: McGraw-Hill Book Company, 1974.

中文参考书目

《辞海》编辑委员会:《辞海》,上海:上海辞书出版社,1979年。

《辞海》(修订稿),上海:上海辞书出版社、上海人民出版社,1976年。

中国企业管理百科全书编辑部:《中国企业管理百科全书》,北京:企业管理出版社,1984年。

马洪主编:《现代中国经济事典》,北京:中国社会科学出版社,1982年。

城乡建设环境保护部政策研究室编:《城乡建设环境保护法规汇编》(第一辑),北京:中国建筑工业出版社,1984年。

同济大学、重庆建筑工程学院、武汉建筑材料工业学院合编:《城市规划原理》,北京:中国建筑工业出版社,1981年。

PINYIN LIST / 汉语拼音对照表

Ānquán cuòshī   安全措施
Àn bǎifēnbǐ jì zūjīn   按百分比计租金
Àn bǎifēnbǐ shōufèi xiéyì   按百分比收费协议
Àn gōngshì bōkuǎn   按公式拨款
Àn mùbiāo zhìdìng guīhuà   按目标制订规划

Bǎifēnbǐ fèiyòng   百分比费用
Bānqiān   搬迁
Bàndúlìshì zhùzhái   半独立式住宅
Bànhuánlù   半环路
Bǎodān   保单
Bǎohù   保护
Bǎojiàn shèshī   保健设施
Bǎozhèngshū   保证书
Bèiyòng biāojià   备用标价
Bèifǔyǎng rénkǒu   被抚养人口
Bìshuì shǒuduàn   避税手段
Biàngēng tōngzhī   变更通知
Biānjiǎodì   边角地
Biànlì shāngdiàn   便利商店
Biānzhì yùsuàn jìhuà xìtǒng   编制预算计划系统
Biāojià zǒngbiǎo   标价总表
Biāozhǔn dàdūshì tǒngjìqū   标准大都市统计区
Biāozhǔn dǐyā dàikuǎn   标准抵押贷款
Bìshuì shǒuduàn   避税手段
Bōdì   拨地
Bōkuǎn   拨款
Bǔchōng wénjiàn   补充文件

Bǔtiē	补贴
Bǔtiē zhùfáng	补贴住房
Bùdòngchǎn	不动产
Bùfú guīdìng shǐyòng	不符规定使用
Bùfú hétóng gōngchéng	不符合同工程
Bùhé biāozhǔn zhùfáng	不合标准住房
Bùjǐngqì dìqū	不景气地区
Bùkěyùjiàn fèi	不可预见费
Bùshòusǔn tiáokuǎn	不受损条款
Bùfèn zhànyòng	部分占用
Cáiliào dìng'é	材料定额
Cáiliào yǔ zhìpǐn jiàndìng	材料与制品鉴定
Cáichǎn bǎoxiǎn	财产保险
Chāiqiānhù	拆迁户
Chāiqiānhù zhùfáng	拆迁户住房
Chǎnquán zhèng	产权证
Chángzhài jījīn	偿债基金
Chǎngqiánqū	厂前区
Cháoxiàng	朝向
Chéngběn chāozhī	成本超支
Chéngběn fēnlèi	成本分类
Chéngběn gūjì	成本估计
Chéngběn hésuàn	成本核算
Chéngběn jiā fèiyòng hétóng	成本加费用合同
Chéngběn jiā fèiyòng zhàngkuǎn	成本加费用账款
Chéngběn jiàgé zhǐshù	成本价格指数
Chéngběn kòngzhì	成本控制
Chéngběn xiàoyì fēnxī	成本效益分析
Chéngjiāo fèiyòng	成交费用
Chéngbāorén	承包人

Chéngbāorénde xuǎnzéquán	承包人的选择权
Chéngnuò	承诺
Chéngzūrén	承租人
Chéngnèi shuāiluòqū	城内衰落区
Chéngqū	城区
Chéngshì biānyuán	城市边缘
Chéngshì dìngjū	城市定居
Chéngshì fángzāi guīhuà	城市防灾规划
Chéngshì fēngmào	城市风貌
Chéngshì fēnqū guīhuà	城市分区规划
Chéngshì fùsū	城市复苏
Chéngshì gǎizào	城市改造
Chéngshì gēngxīn	城市更新
Chéngshì guīhuà	城市规划
Chéngshì guīhuà guǎnlǐjú	城市规划管理局
Chéngshì guīhuà wěiyuánhuì	城市规划委员会
Chéngshì guīmó	城市规模
Chéngshì jīngjìxué	城市经济学
Chéngshì jūmín	城市居民
Chéngshì kāifā huódòng zèngkuǎn	城市开发活动赠款
Chéngshì kàngzhèn guīhuà	城市抗震规划
Chéngshì měihuà yùndòng	城市美化运动
Chéngshì rénfáng guīhuà	城市人防规划
Chéngshì shèhuìxué	城市社会学
Chéngshì shèjì	城市设计
Chéngshì shēngtàixué	城市生态学
Chéngshì wàiwéi	城市外围
Chéngshì wǎngjià	城市网架
Chéngshì wújìhuà kuòzhǎn	城市无计划扩展
Chéngshì xíngtài	城市形态

Chéngshì xìngzhì	城市性质
Chéngshì yòngdì píngjià	城市用地评价
Chéngshì yuǎnjiāo	城市远郊
Chéngshì zhèngcè	城市政策
Chéngfá hé jiǎnglì tiáokuǎn	惩罚和奖励条款
Chíxù nénglì	持续能力
Chǐcùn xiétiáo	尺寸协调
Chóngchóu zījīn	重筹资金
Chóngjiàn	重建
Chóngxīn qūhuà	重新区划
Chóngzhì chéngběn	重置成本
Chóucuò jiànzào zījīn	筹措建造资金
Chóujiàn xiàngmù	筹建项目
Chóuzī	筹资
Chūjià	出价
Chūrùquán	出入权
Chūshòu zūhuí	出售租回
Chūzūrén	出租人
Chūbù shèjì	初步设计
Chūjí chéngshì	初级城市
Chūjí dǐyā dàikuǎn shìchǎng	初级抵押贷款市场
Chǔxù dàikuǎn xiéhuì	储蓄贷款协会
Chuántǒng jiànzhù	传统建筑
Cūlüègūsuàn	粗略估算
Cuòtáishì zhùzhái	错台式住宅
Dàbǎnshì zhùfáng	大板式住房
Dàchéngshì	大城市
Dàdūshì	大都市
Dàdūshì dìqū	大都市地区
Dàdūshì tǒngjìqū	大都市统计区
Dàjiātíng	大家庭

Dàliàng kèyùn	大量客运
Dàihuílángde xiǎozhùzhái	带回廊的小住宅
Dàixíng shāngyèqū	带形商业区
Dàikuǎn yǔ jiàzhíbǐ	贷款与价值比
Dānbǎopǐn	担保品
Dānfāng chéngběn	单方成本
Dānfāng yònggōng	单方用工
Dānfāng yòngliào	单方用料
Dānjià	单价
Dānshēn sùshè	单身宿舍
Dānwèi chéngběn	单位成本
Dānwèi zhùfáng	单位住房
Dīcéng gāomìdù	低层高密度
Dīcéng jiànzhù	低层建筑
Dǐyā dàikuǎn	抵押贷款
Dǐyā dàikuǎn yínháng	抵押贷款银行
Dǐyā shōuyì zhàiquàn	抵押收益债券
Dìchǎn chǔbèi	地产储备
Dìduàntú	地段图
Dìfāng nénglì	地方能力
Dìfāng xiàngmù	地方项目
Dìfāng zhèngfǔ	地方政府
Dìjícè	地籍册
Dìjià	地价
Dìjiè	地界
Dìjiè cèliáng	地界测量
Dìjiè língxiàn	地界零线
Dìkuài	地块
Dìqì	地契
Dìqū	地区
Dìxià jīngjì	地下经济
Dìxíng cèliáng	地形测量

Dìyìquán	地役权
Dìzēng chánghuán dǐyā dàikuǎn	递增偿还抵押贷款
Dìng'é	定额
Dìngjīn	定金
Dìngjū gēngzhòng	定居耕种
Dìngshí fēnxiǎng ānpái	"定时分享"安排
Dúlìfèi	独立费
Dúlìshì dúhù zhùzhái	独立式独户住宅
Dúlìshì zhùzhái	独立式住宅
Dúlì xīnchéng	独立新城
Dúmén dúhù	独门独户
Dúshēng zǐnǚ hù	独生子女户
Duìjǐng	对景
Duōcéng chūzū chǎngfáng	多层出租厂房
Duōhù zhùfáng	多户住房
Duōyàngxìng	多样性
Duōyòng qūyù	多用区域
Duōyòngtú jiànshè	多用途建设
Duōyòngtú qūhuà	多用途区划
Duōyuánhuà shèhuì	多元化社会
Duōzhōngxīn chéngshì	多中心城市
Èrjí dǐyā dàikuǎn shìchǎng	二级抵押贷款市场
Èrliánshì zhùzhái	二联式住宅
Fābāo	发包
Fǎdìng lìrùn	法定利润
Fǎlìng	法令
Fānxīn	翻新
Fānxiū	翻修
Fǎnkuì	反馈

Fànlàn píngyuán	泛滥平原
Fāng'àn shèjì	方案设计
Fānggéshì lùwǎng	方格式路网
Fánghùdài	防护带
Fángyùxìng kōngjiān	防御性空间
Fángdìchǎn	房地产
Fángdìchǎn kèshuì jiàzhí	房地产课税价值
Fángguǎnbùmén zhùfáng	房管部门住房
Fángguǎnsuǒ	房管所
Fángzū kòngzhì	房租控制
Fàngkuān tiáolì	放宽条例
Fàngqì chǎnquán	放弃产权
Fēidì	飞地
Fēishēngchǎnxìng jiànshè	非生产性建设
Fēizhèngguī hángyè	非正规行业
Fēizhèngfǔ zǔzhī	非政府组织
Fèiqì	废弃
Fèiwù guǎnlǐ	废物管理
Fèiwù lìyòng	废物利用
Fèi	费
Fēnbāo hétóng	分包合同
Fēnbāorén	分包人
Fēnbù gōngchéng	分部工程
Fēncéng chūzū chǎngfáng	分层出租厂房
Fēnhù	分户
Fēnjí	分级
Fēnkuài tǔdì	分块土地
Fēnqī chánghuán	分期偿还
Fēnqī tíkuǎn	分期提款
Fēnqū zhuānyòng	分区专用
Fēnxiàng gōngchéng	分项工程
Fēnxiǎng zēngzhí dǐyā dàikuǎn	分享增值抵押贷款

Fēnxiànggōngchéng yànshōu	分项工程验收
Fēngjǐng jiànzhùxué	风景建筑学
Fēngjǐng míngshèng bǎohù	风景名胜保护
Fēngxiǎn dānbǎo	风险担保
Fēngxiǎn fēnxī	风险分析
Fúwù hángyè	服务行业
Fúwù rénkǒu	服务人口
Fúdòngqū	浮动区
Fúhé shǐyòngzhèngmíngshū	符合使用证明书
Fǔzhù dàolù	辅助道路
Fǔzhù miànjī	辅助面积
Fùkuǎn shēnqǐngshū	付款申请书
Fùkuǎn zhèngmíngshū	付款证明书
Fùzhài	负债
Fùdài yòngtú	附带用途
Fùshè fúwù	附设服务
Fùshǔ jiànzhù	附属建筑
Fùxīng	复兴
Fùyè shēngchǎn	副业生产
Fùzuòyòng	副作用
Gǎijiàn gōngchéng	改建工程
Gǎizào	改造
Gàiniàn fāzhǎn	概念发展
Gàisuàn dìng'é	概算定额
Gǎijìnhòu jìngzēngzhí	改进后净增值
Gàiyàoxìng guīhuàtú	概要性规划图
Gànxiàn	干线
Gànggǎn zuòyòng	杠杆作用
Gāocéng dīmìdù	高层低密度
Gāocéng gāomìdù	高层高密度
Gāocéng jiànzhù	高层建筑

Gāosù gōnglù	高速公路
Gémìng jìniàndì	革命纪念地
Gèbiédiǎn qūhuà	个别点区划
Gètǐhù	个体户
Gēngxīn	更新
Gēngxīn gǎizào cuòshī tóuzī	更新改造措施投资
Gēngxīn gǎizào cuòshī xiàngmù	更新改造措施项目
Gōngchǎng yùzhì zhùfáng	工厂预制住房
Gōngchéng biàngēng	工程变更
Gōngchéng biàngēng tōngzhīdān	工程变更通知单
Gōngchéngchù (gōngqū)	工程处（工区）
Gōngchéngduì (gōngduàn)	工程队（工段）
Gōngchéng fēnxiàng jiàzhíbiǎo	工程分项价值表
Gōngchéng jìndùkuǎn	工程进度款
Gōngchéng rènwùdān	工程任务单
Gōngchéngshī rènkě zhèngmíng	工程师认可证明
Gōngchéng xíngxiàng jìndù	工程形象进度
Gōngchéng yòngliào yǔ shèbèi qīngdān	工程用料与设备清单
Gōngchéng yùsuàn	工程预算
Gōngchéng zhǐhuībù	工程指挥部
Gōngchéng zhǔrèn	工程主任
Gōngduànzhǎng	工段长
Gōngnóng liánhé qǐyè	工农联合企业
Gōngnuǎn, tōngfēng, kōngtiáo	供暖、通风、空调
Gōngyèhuà jiànzhù tǐxì	工业化建筑体系
Gōngyèhuà zhùfáng	工业化住房

Gōngyè jiànzhù	工业建筑
Gōngyèqū	工业区
Gōngyèyuán	工业园
Gōngyè zhàiquàn	工业债券
Gōngzhǎng	工长
Gōnggòng gōngchéng xiàng-mù tóuzī	公共工程项目投资
Gōnggòng guānxì	公共关系
Gōnggòng jīgòu	公共机构
Gōnggòng qìchē zhīxiàn xìtǒng	公共汽车支线系统
Gōnggòng shèshī	公共设施
Gōnggòngshèshīde gǎijìn	公共设施的改进
Gōnggòng zhùfáng	公共住房
Gōngjiàn mínzhù	公建民助
Gōngpíng shìchǎng jiàzhí	公平市场价值
Gōngpíng zhùfáng zhèngcè	公平住房政策
Gōngsī	公司
Gōngsī hézuò	公私合作
Gōngyòngbùfènde wéihù	公用部分的维护
Gōngyòng shèshī	公用设施
Gōngyòng shèshī diàochá	公用设施调查
Gōngyòng shèshī fèiyòng	公用设施费用
Gōngyǒu bùfèn	公有部分
Gōngyǒu fángdìchǎn zhuǎnràng	公有房地产转让
Gōngyù	公寓
Gōngyùlóu	公寓楼
Gōngzhòng cānyù	公众参与
Gōngzhòng yìjiàn tīngqǔhuì	公众意见听取会
Gōngzhù zhùfáng	公助住房
Gōngnéng biāozhǔn qūhuà	功能标准区划

Gōngnéng fǎguī	功能法规
Gōngnéng fēnqū	功能分区
Gōngnéng guīhuà	功能规划
Gōngnéngxìng guòshí	功能性过时
Gōngshànsùde sījiā zhùzhái	供膳宿的私家住宅
Gōngxuǎnzé fāng'àn	供选择方案
Gòngchóu zījīn	共筹资金
Gòngtóng bǎoxiǎn	共同保险
Gòuwù shāngchǎng	购物商场
Gòuwù zhōngxīn	购物中心
Gūjià	估价
Gǔjiànzhù bǎohù	古建筑保护
Gǔlìxìng qūhuà	鼓励性区划
Gùdìng jiàgé hétóng	固定价格合同
Gùdìng jiàgé jìngzhēng	固定价格竞争
Gùdìnglìlǜ dǐyā dàikuǎn	固定利率抵押贷款
Gùdìng zīběn tóuzī	固定资本投资
Gùdìng zīchǎn tóuzī	固定资产投资
Gùyǒu quánlì	固有权利
Gùwèn	顾问
Gùwèn gōngchéngshī	顾问工程师
Guānjiàn xiànlùfǎ	关键线路法
Guānjiàn zūhù	关键租户
Guǎnlǐfèi	管理费
Guǎnlǐ jiégòu	管理结构
Guǎnwǎng zōnghé	管网综合
Guīchéng	规程
Guīdìng	规定
Guīgé fǎguī	规格法规
Guīhuà wěiyuánhuì	规划委员会
Guīhuà xiétiáo wěiyuánhuì	规划协调委员会
Guīhuà zǒngtú	规划总图

Guójiā tóuzī	国家投资
Guójiā xiàngmù (bùzhíshǔ xiàngmù)	国家项目（部直属项目）
Guójiā zhēngdìquán	国家征地权
Guómín jīngjì hé shèhuì fāzhǎn jìhuà	国民经济和社会发展计划
Guónèi dàikuǎn	国内贷款
Guótǔ guīhuà	国土规划
Guówài dàikuǎn	国外贷款
Guòdùqū	过渡区
Guòhù	过户
Guòshí	过时
Guòshí zhùfáng	过时住房
Hángkōng cèliáng	航空测量
Hébìng tǔdì	合并土地
Héchéng	合乘
Hégé zhèngmíngshū	合格证明书
Héhuǒ chéngchē	合伙乘车
Hétóng	合同
Hétóng dānbǎoshū	合同担保书
Hétóng guǎnlǐ	合同管理
Hétóng wénjiàn	合同文件
Hétóng xiàngmù	合同项目
Hézhù	合住
Hézhù zhùfáng	合住住房
Hézī jīngyíng	合资经营
Hézū zhùzhái	合租住宅
Héxīn jiātíng (xiǎo jiātíng)	核心家庭（小家庭）
Héxīn zhùfáng	核心住房
Hóngxiàn zhùxiāo	红线注销
Hútòng	胡同

Hù	户
Hùjí	户籍
Hùkǒu	户口
Hùkǒubù	户口簿
Hùkǒu dēngjì	户口登记
Hùshìbǐ	户室比
Hùxíng	户型
Hùxíngbǐ	户型比
Hùzhù chǔxù yínháng	互助储蓄银行
Hùzhù zhùfáng xiéhuì	互助住房协会
Huāyuán gōngyù	花园公寓
Huàshì gōngyù	画室公寓
Huánchéng gōnglù	环城公路
Huánjìng biāozhǔn	环境标准
Huánjìng shèjì	环境设计
Huánjìng yǐngxiǎng bàogào	环境影响报告
Hùnhé jīngjì	混合经济
Hùnhé shīgōngduì	混合施工队
Huódòng fēnxī	活动分析
Huódòng zhùfáng	活动住房
Jiāsù zhéjiù	加速折旧
Jīhuì chéngběn	机会成本
Jīběn biāojià	基本标价
Jīběn jiànshè	基本建设
Jīběn jiànshè tóuzī	基本建设投资
Jīběn jiànshè xiàngmù	基本建设项目
Jīběn jīngjì	基本经济
Jīběn rénkǒu	基本人口
Jīběn zūjīn	基本租金
Jīchǔ shèshī	基础设施
Jīdǐmiànjī	基底面积

Jīzhǔn shùjù	基准数据
Jíhé chéngshì	集合城市
Jíhé gōngyù	集合公寓
Jìhuà jiēduàn	计划阶段
Jìhuà píngshěn fǎ	计划评审法
Jìshù píngdìng	技术评定
Jìshù shèjì	技术设计
Jìshùxìng jiàndìng	技术性鉴定
Jìshù zhuǎnràng	技术转让
Jìcún dānbǎopǐn	寄存担保品
Jiājiàn gōngchéng	加建工程
Jiǎncháyuán	检查员
Jiǎnjì chéngběn	"减记成本"
Jiǎnlòu dājiàn	简陋搭建
Jiànjiēfèi	间接费
Jiànchéng huánjìng	建成环境
Jiànchéngqū	建成区
Jiànchéng tóurù shēngchǎn xiàngmù	建成投入生产项目
Jiànfáng xiéhuì	建房协会
Jiànshè	建设
Jiànshè chǎngdì fēnxī	建设场地分析
Jiànshè chǎngdì guīhuà	建设场地规划
Jiànshè chǎngdì kāifā	建设场地开发
Jiànfèng chāzhēn	"见缝插针"
Jiàngdī mìdù qūhuà	降低密度区划
Jiǎngfá tiáokuǎn	奖罚条款
Jiànshè dānwèi	建设单位
Jiànshè quēxiàn qīngdān	建设缺陷清单
Jiànshè xiàngmù	建设项目
Jiànshè xiàngmù guīmó	建设项目规模
Jiànshè xiàngmù tóuzī yùsuàn	建设项目投资预算

Jiànshè xiàngmù yèzhǔ	建设项目业主
Jiànyì guīhuà	建议规划
Jiànzàoliàng	建造量
Jiànzhù ānzhuāng gōngchéng	建筑安装工程
Jiànzhù ānzhuāng gōngchéngliàng	建筑安装工程量
Jiànzhù ānzhuāng gōngzuòliàng	建筑安装工作量
Jiànzhù biāozhǔn	建筑标准
Jiànzhù bùzhìtú	建筑布置图
Jiànzhù cáiliào jiǎnyàn	建筑材料检验
Jiànzhù chéngběn	建筑成本
Jiànzhù fǎguī yǔ biāozhǔn	建筑法规与标准
Jiànzhù gāodù	建筑高度
Jiànzhù gāodù qūhuà	建筑高度区划
Jiànzhù gōngsī	建筑公司
Jiànzhù gōngzhǒng	建筑工种
Jiànzhù gòuzào guīgé	建筑构造规格
Jiànzhù hóngxiàn	建筑红线
Jiànzhù jiāncháyuán	建筑监察员
Jiànzhù máomiànjī	建筑毛面积
Jiànzhù miànjī	建筑面积
Jiànzhù miànjī zhǐbiāo	建筑面积指标
Jiànzhù miànjībǐ	建筑面积比
Jiànzhù mìdù	建筑密度
Jiànzhù xìbù shuōmíng	建筑细部说明
Jiànzhù zàojià	建筑造价
Jiànzhùfǎ	建筑法
Jiànzhùqún	建筑群
Jiànzhùshī gōngchéngshī shìwùsuǒ	建筑师工程师事务所

Jiànzhùshī rènkězhèngmíng	建筑师认可证明
Jiànzhùshī、Jiànzhùshī shìwùsuǒ	建筑师、建筑师事务所
Jiànzhùxué	建筑学
Jiànzhùyèwù	建筑业务
Jiǎonǎozhīhuì	绞脑汁会
Jiāoqū	郊区
Jiāotōng gōngjù fēnlèi	交通工具分类
Jiāotōng guǎnlǐ	交通管理
Jiāotōng guǎnzhìduàn	交通管制段
Jiāotōng yùcè	交通预测
Jiāoyàoshi gōngchéng	交钥匙工程
Jiāshǔ sùshè	家属宿舍
Jiātíng	家庭
Jiātíng gōngyè	家庭工业
Jiātíng qiānyí	家庭迁移
Jiàzhí gōngchéng	价值工程
Jiēdào bànshìchù	街道办事处
Jiēdào gōngchǎng	街道工厂
Jiēdào xiǎopǐn	街道小品
Jiégòuxìng shīyè	结构性失业
Jièjǐng	借景
Jiējǐng	街景
Jièkuǎn tiáojiàn	借款条件
Jiésuàn	结算
Jiésuàn fèiyòng	结算费用
Jiéyuē yòngdì	节约用地
Jìnduānlù	尽端路
Jīngjì gōngyù	经济公寓
Jīngjì tèqū	经济特区
Jīngjìqū	经济区
Jīngjìxìng táotài	经济性淘汰

Jìngshōuyìlán	净收益栏
Jīngyíng yùsuàn	经营预算
Jìngzhēngxìng bōkuǎn	竞争性拨款
Jìnjiāoqū	近郊区
Jíshì	集市
Jiùyè zhuǎnyí	就业转移
Júbù guīhuà	局部规划
Juécè guòchéng	决策过程
Juécè zīliào	决策资料
Jùjítǐ	聚集体
Jūmín liánfáng	居民联防
Jūmín wěiyuánhuì	居民委员会
Jūmín xiǎozǔ	居民小组
Jùngōng yànshōu	竣工验收
Jùngōngtú	竣工图
Jūshì	居室
Jūzhù bùfāngbiàn hù	居住不方便户
Jūzhù dānyuán	居住单元
Jūzhù jiànzhù	居住建筑
Jūzhù jiànzhù miànjī	居住建筑面积
Jūzhù jiànzhù miànjī mìdù	居住建筑面积密度
Jūzhù jìngmìdù	居住净密度
Jūzhù máomìdù	居住毛密度
Jūzhù miànjī	居住面积
Jūzhù miànjī mìdù	居住面积密度
Jūzhù mìdù	居住密度
Jūzhù rénkǒu mìdù	居住人口密度
Jūzhù xiǎoqū	居住小区
Jūzhù yōngjǐhù	居住拥挤户
Jūzhùqū	居住区
Jūzhùquán	居住权
Jūzhùyòngdì jìngmiànjī	居住用地净面积

Kāifā jiǎnglì	开发奖励
Kāifā、jiànshè	开发、建设
Kāifāquán	开发权
Kāifāzhě、jiànshèzhě	开发者、建设者
Kěbiàn lìlǜ dǐyā dàikuǎn	可变利率抵押贷款
Kěchūzū jìngmiànjī	可出租净面积
Kěchūzū máomiànjī	可出租毛面积
Kěchūzū miànjī	可出租面积
Kěchūzū miànjībǐ	可出租面积比
Kějiàn miànjī	可建面积
Kětiáo lìlǜ dǐyā dàikuǎn	可调利率抵押贷款
Kěxíngxìng yánjiū	可行性研究
Kězhuǎnràng kāifāquán	可转让开发权
Kèhù	客户
Kòngdì	空地
Kòngdì huò kōngfáng diàochábiǎo	空地或空房调查表
Kōngfáng lǜ	空房率
Kōngjiān bùjú	空间布局
Kǒudài gōngyuán (Xiǎoyóuyuán)	"口袋公园"（小游园）
Kuàisù gàndào	快速干道
Kuàisù shīgōngfǎ	快速施工法
Kuàisù yùnshū	快速运输
Kuòdà nénglì	扩大能力
Lǎnzǒng dǐyā	揽总抵押
Lǎnzǒng dǐyā xìntuōshū	揽总抵押信托书
Láodòng dìng'é	劳动定额
Láodòng pínghéng	劳动平衡
Láodòng rénkǒu	劳动人口
Lǎonián zhùfáng	老年住房

Lìshǐ wénhuà míngchéng bǎohù	历史文化名城保护
Lìshǐxìng jiànzhù bǎohù	历史性建筑保护
Lìshǐxìng jiànzhù bǎohù zhèngmíng	历史性建筑保护证明
Lìshǐxìng shìqū	历史性市区
Lìtǐ jiāochā	立体交叉
Lìrùnlǜ	利润率
Lìyòng xìshù	利用系数
Liánlìshì zhùzhái	联立式住宅
Liánghǎo jīngjì xiàoyì	良好经济效益
Línlǐ	邻里
Línlǐ bǎohù	邻里保护
Línlǐ cānyù	邻里参与
Línlǐ fùsū	邻里复苏
Línlǐ zìwèi	邻里自卫
Línyīn gōnglù	林荫公路
Línjiēmiàn	临街面
Línjiè fēnxī (ménkǎn lǐlùn)	临界分析（门槛理论）
Línshí shèshī	临时设施
Língxīng jiànzhù	零星建筑
Liúzhìquán	留置权
Liúshuǐxiàn	流水线
Liúshuǐ zuòyèfǎ	流水作业法
Lùjiān	路肩
Lǚxíng hétóng bǎodān	履行合同保单
Lǜdài	绿带
Lǜdì	绿地
Mǎimàiquán	买卖权
Màiqì	卖契
Měiguān kòngzhì	美观控制

Měiguān qūhuà	美观区划
Měiguó lìshǐxìng jiànzhù diàochá	美国历史性建筑调查
Měihuà	美化
Mìdù tiáojì	密度调剂
Miànjī gūjiàfǎ	面积估价法
Mínjū	民居
Mínyòng jiànzhù	民用建筑
Móshù	模数
Móshù xiétiáo	模数协调
Móshùzhì zhùfáng	模数制住房
Mùjià fángwū	木架房屋
Nàoshìqū	闹市区
Nèitíngshì jiànzhù	内庭式建筑
Néngyuán bǎohù	能源保护
Néngyuán xiàolǜbǐ	能源效率比
Néngliàng xiàolǜ biāozhǔn	能量效率标准
Niánlíng fēnbù	年龄分布
Nónggōngshāng liánhéqǐyè	农工商联合企业
Nóngtián qīngmiáo bǔcháng	农田青苗补偿
Pàichūsuǒ	派出所
Pèitào xiàngmù	配套项目
Pílián zhùzhái	毗连住宅
Pínkùn chéngshì	贫困城市
Pínkùnxiàn	贫困线
Pínmínkū	贫民窟
Píngjiāo	平交
Píngjūn měihù rénkǒu	平均每户人口
Píngjūn měirén jūzhù miànjī	平均每人居住面积
Píngjūn měitào jiànzhù miànjī	平均每套建筑面积

汉语拼音对照表

Píngzhěng tǔdì	平整土地
Pòluò	破落
Qīshēnchù	栖身处
Qǐzhōngdiǎn diàochá	起终点调查
Qìcái diàodùyuán	器材调度员
Qīnzhàn	侵占
Qū	区
Qū fángguǎnjú	区房管局
Qūhuà	区划
Qūhuà shēnsù wěiyuánhuì	区划申诉委员会
Qūhuà shìqū	区划市区
Qūhuà shòuquán	区划授权
Qūhuà tèxǔzhèng	区划特许证
Qūhuà xíngzhèngguān	区划行政官
Qūhuà xiūzhèng	区划修正
Qūhuà xǔkězhèng	区划许可证
Qūyù gōngrè	区域供热
Qūyù guīhuà wěiyuánhuì	区域规划委员会
Qǔfèi biāozhǔn	取费标准
Qǔxiāo shúhuíquán	取消赎回权
Quánlì fēnsàn	权力分散
Quánbù jùngōng xiàngmù	全部竣工项目
Quánshòumìng fèiyòng jìsuàn	全寿命费用计算
Quèrèn	确认
Ràoxínglù	绕行路
Rénkǒu fēnbù	人口分布
Rénkǒu jīxiè zēngzhǎng	人口机械增长
Rénkǒu jiǎnshǎo	人口减少
Rénkǒu liúdòng	人口流动

Rénkǒu mìdù	人口密度
Rénkǒu pǔchá jiēqū	人口普查街区
Rénkǒu pǔchá qūduàn	人口普查区段
Rénkǒu tǒngjìxué	人口统计学
Rénkǒu zìrán zēngzhǎng	人口自然增长
Rénlèi chǐdù	人类尺度
Rénlèi fúwù	人类服务
Rénlèi qúnjūxué	人类群居学
Rénlèi shēngtàixué	人类生态学
Rénlì zīyuán	人力资源
Rénxíng línyīnlù	人行林荫路
Rìtuō zhōngxīn	日托中心
Rìzhào biāozhǔn	日照标准
Róngjīlǜ	容积率
Sānliánshì zhùzhái	三联式住宅
Shāngyè miànjī	商业面积
Shāngyè qū	商业区
Shāngyè wǎngdiǎn	商业网点
Shāngyè zhōngxīnqū	商业中心区
Shàngcéng guǎnlǐ	上层管理
Shàngkōng shǐyòngquán	上空使用权
Shǎoshùrénzhǒng jùjūqū	少数人种聚居区
Shèjì dānwèi	设计单位
Shèjì gàisuàn	设计概算
Shèjì gàiyào	设计概要
Shèjì jiān shīgōng	设计兼施工
Shèjì jiēduàn	设计阶段
Shèjì rènwùshū	设计任务书
Shèjì yùsuàn	设计预算
Shèjì zǔzhǎng	设计组长
Shèhuì cānyù	社会参与

Shèhuì fúwù	社会服务
Shèhuì guīhuà	社会规划
Shèhuì jiégòu	社会结构
Shèhuì jiēcéng huàfēn	社会阶层划分
Shèhuì jīngjì zōnghé zhǐbiāo	社会经济综合指标
Shèhuì yītǐhuà	社会一体化
Shèhuì zhǐshù	社会指数
Shèqū	社区
Shèqū cānyù	社区参与
Shèqū dàxué	社区大学
Shèqū jiànshè	社区建设
Shèqū kāifā gōngsī	社区开发公司
Shèqū shèshī	社区设施
Shèqū xiéhuì	社区协会
Shèqū zhōngxīn	社区中心
Shètuán	社团
Shēngchǎn guǎnlǐ	生产管理
Shēngchǎn jīdì	生产基地
Shēngchǎnlǜ	生产率
Shēngchǎnxìng jiànshè	生产性建设
Shēnghuó fāngshì	生活方式
Shēnghuó fúwù shèshī	生活服务设施
Shēnghuó jīdì	生活基地
Shēngtàixué	生态学
Shēngzhǎngdì	生长地
Shěng	省
Shěngxiáshì	省辖市
Shīgōng	施工
Shīgōng dānwèi	施工单位
Shīgōng dìng'é	施工定额
Shīgōngduì	施工队

Shīgōng duìwù	施工队伍
Shīgōng gōngqī	施工工期
Shīgōng guǎnlǐ	施工管理
Shīgōng guīfàn yánjiūyuàn	施工规范研究院
Shīgōng jiāndū	施工监督
Shīgōng jiǎncháyuán	施工检查员
Shīgōng jiēduàn	施工阶段
Shīgōng jìndù jìhuà	施工进度计划
Shīgōng Jīnglǐ	施工经理
Shīgōngtú	施工图
Shīgōngtú shèjì	施工图设计
Shīgōngtú yùsuàn	施工图预算
Shīgōng wénjiàn	施工文件
Shīgōng xiàngmù	施工项目
Shīgōng yùsuàn	施工预算
Shīgōng zǒngpíngmiàntú	施工总平面图
Shīgōngzǔzhī shèjì	施工组织设计
Shīgōng zuòyè jìhuà	施工作业计划
Shíxiào	时效
Shíjì xūqiú	实际需求
Shítǐ guīhuà	实体规划
Shíyòng yuánlín bùzhì	食用园林布置
Shǐyònghòu diàochá	使用后调查
Shǐyòng miànjī	使用面积
Shìfàn gōngchéng	示范工程
Shì	市
Shìchǎng fēnxī	市场分析
Shìchǎng jiàgé	市场价格
Shìchǎnglìlǜ dàikuǎn	市场利率贷款
Shìchǎng suǒzàidì	市场所在地
Shìchǎng zūjià	市场租价
Shì fángguǎnjú	市房管局

Shìguǎnxiàn	市管县
Shìmíncānyù	市民参与
Shìqū	市区
Shìxiáxiàn	市辖县
Shì xíngzhèngguān	市行政官
Shìzhèn liánlìshì zhùzhái	市镇联立式住宅
Shìzhèng gōngyòng shèshī	市政公用设施
Shìzhèng jiànzhù	市政建筑
Shìzhèng zhōngxīn	市政中心
Shìdiǎn xiàngmù	试点项目
Shíyànshì zīgé zhèngmíng	实验室资格证明
Shìnèi shèjì	室内设计
Shìyìngxìng zàishǐyòng	适应性再使用
Shìyìngzìránde shèjì	适应自然的设计
Shìyòng jìshù	适用技术
Shōurù cáizhèng	收入财政
Shōurù liúliàng	收入流量
Shōurù zēngzhí zuòyòng	收入增殖作用
Shōuzhī pínghéngdiǎn	收支平衡点
Shǒucì fùkuǎn	首次付款
Shòuzīzhù chéngshì	受资助城市
Shùxiàng guīhuà	竖向规划
Shùliàng míngxìbiǎo	数量明细表
Shuāiluò	衰落
Shuāituì línlǐ	衰退邻里
Shuōmíngshū	说明书
Sībìxìng	私蔽性
Sīyǒu bùfèn	私有部分
Sǐhútòng	死胡同
Sìhéyuàn	四合院
Sìliánshì zhùzhái	四联式住宅
Suōjìn	缩进

Tángdì	塘地
Tàofáng	套房
Tàolóufáng	套楼房
Tèbié gōngzuòzǔ	特别工作组
Tèdà wěishù dǐyā dàikuǎn	特大尾数抵押贷款
Tèqū	特区
Tígāo zhìliàng	提高质量
Tígōng chǎngdì yǔ jīchǔ shèshī	提供场地与基础设施
Tígōng jiùyè	提供就业
Tǐjīguīdìng	体积规定
Tìhuàn jìshù	替换技术
Tiānjìxiàn	天际线
Tiánkòngxìng jiànshè	填空性建设
Tiáokuǎn	条款
Tiáoxíng shāngyèqū	条形商业区
Tiáozhěng tiáokuǎn	调整条款
Tíngjiàn huǎnjiàn xiàngmù	停建缓建项目
Tōngqínzhě	通勤者
Tōngxíngquán	通行权
Tōngyòng jiànzhù tǐxì	通用建筑体系
Tóngxīnqū lǐlùn	同心区理论
Tóngyīxìng	同一性
Tǒngjiàn zhùfáng	统建住房
Tǒngyī jiànzhù zhǐbiāo	统一建筑指标
Tóubiāo	投标
Tóubiāo bǎodān	投标保单
Tóubiāo jiàgé	投标价格
Tóubiāo wénjiàn yājīn	投标文件押金
Tóujī	投机
Tónzī jiǎnshuì'é	投资减税额
Tūpò xíngdòng	突破行动

Tǔdì bǎohù	土地保护
Tǔdì bǔchángfèi	土地补偿费
Tǔdì cèliáng	土地测量
Tǔdì chíyǒufèi	土地持有费
Tǔdì fēnpèi	土地分配
Tǔdì huò zījīn fēnpèi	土地或资金分配
Tǔdì lìyòng diàochá	土地利用调查
Tǔdì lìyòng guīhuàtú	土地利用规划图
Tǔdì shǐyòng kòngzhì	土地使用控制
Tǔdì tiáozhěng	土地调整
Tǔdì zhànyòng quán	土地占用权
Tǔdì zhěngzhì	土地整治
Tǔdì zūyuē	土地租约
Tǔrǎng diàochá	土壤调查
Tuōqiàn dǐyā dàikuǎn	拖欠抵押贷款
Wàizài fùzuòyòng	外在副作用
Wángōng bǎozhèngshū	完工保证书
Wǎnqī bìngrén hùlǐyuàn	晚期病人护理院
Wànyuán dìng'é	万元定额
Wǎnggé guīhuàtú	网格规划图
Wǎngluò fēnxī	网络分析
Wēifáng	危房
Wéiyuē	违约
Wéizhāng jiànzhù	违章建筑
Wéizhāng jūliúdì	违章居留地
Wéixiū bǎozhèngshū	维修保证书
Wèixīngchéng	卫星城
Wèizūhùde fúwù	为租户的服务
Wèigǎishàn tǔdì	未改善土地
Wèizhì lǐlùn	位置理论
Wénhuà tèsèqū shèjì	文化特色区设计

Wénjiàoqū	文教区
Wénwùgǔjì bǎohù	文物古迹保护
Wòchéng jiāoqū	卧城郊区
Wòchéngqū	卧城区
Wūshuǐ chǔlǐ xìtǒng	污水处理系统
Wūshuǐ èrjí chǔlǐ	污水二级处理
Wūshuǐ shēndù chǔlǐ	污水深度处理
Wūshuǐ yījí chǔlǐ	污水一级处理
Wúfánghù	无房户
Wúzhàng'ài huánjìng	无障碍环境
Wúzǐnǚhù	无子女户
Xīyǐn dìqū	吸引地区
Xìtǒng fāngfǎ	系统方法
Xìtǒng fēnxī	系统分析
Xìtǒng guǎnlǐ	系统管理
Xìbāoxíng zēngzhǎng	细胞型增长
Xiàn	县
Xiàn xíngzhèngguān	县行政官
Xiànchǎng diàochá	现场调查
Xiànchǎng gōngchéngshī	现场工程师
Xiànchǎng jiāndū	现场监督
Xiànchǎng jiànzhùshī	现场建筑师
Xiànjīn liúliàng	现金流量
Xiànyǒu zhùfángliàng	现有住房量
Xiāng	乡
Xiāngzhèn qǐyè	乡镇企业
Xiāngguān jīgòuwǎng	相关机构网
Xiāngyìng zèngkuǎn	相应赠款
Xiángxì guīhuà	详细规划
Xiàngxīnhuà	向心化
Xiāochú zhǒngzú gélí	消除种族隔离

Xiǎochéngshì	小城市
Xiǎochúfáng	小厨房
Xiǎoqìhòu	小气候
Xiǎoxíng qǐyè	小型企业
Xīnchéng	新城
Xìnxī kēxué	信息科学
Xìnyòng zhàiquàn	信用债券
Xīngyèqū	兴业区
Xiūfù	修复
Xǔkě shǐyòng	许可使用
Xùshuǐcéng	蓄水层
Xuǎnzéxìng bōkuǎn	选择性拨款
Yǎntǔ zhùfáng	掩土住房
Yǎnbiànzhōngde línlǐ	演变中的邻里
Yàngbǎn fǎguī	样板法规
Yáodòng zhùzhái	窑洞住宅
Yáogǎn jìshù	遥感技术
Yèzhǔ	业主
Yīlǎnzi jiāoyì	一揽子交易
Yīlǎnzi zèngkuǎn	一揽子赠款
Yírén shìwù	宜人事物
Yìjìng	意境
Yǐnbìgōngchéng yànshōu	隐蔽工程验收
Yǐngxiǎng	影响
Yìngjí zhùfáng	应急住房
Yìngfèiyòng	硬费用
Yòngdì pínghéng	用地平衡
Yònghù diàochá	用户调查
Yònghùfèi	用户费
Yònghù xūqiú	用户需求
Yōuhuì lìlǜ	优惠利率

Yōuxiān chēdào	优先车道
Yóukè róngliàng	游客容量
Yǒuguīhuàde dìduàn jiànshè	有规划的地段建设
Yǒuhài fèiwù	有害废物
Yǒuhài gōngyè	有害工业
Yǒushèshī yòngdì	有设施用地
Yùcè	预测
Yùfù cáiliàokuǎn	预付材料款
Yùsuàn dìng'é	预算定额
Yuánchāi yuánjiàn	原拆原建
Yuǎnjiāoqū	远郊区
Yuècéng gōngyù	跃层公寓
Yùnchóuxué	运筹学
Záhuòdiàn	杂货店
Zàilìyòng	再利用
Zànkòukuǎn	暂扣款
Zàoshēng kòngzhì	噪声控制
Zēngzhǎnglǜ	增长率
Zēngzhǎng zhōngxīn	增长中心
Zēngzhí	增值
Zèngkuǎn	赠款
Zháijīdì	宅基地
Zhàiquàn	债券
Zhàiwù qīngcháng fúwù	债务清偿服务
Zhànyòng	占用
Zhànyòngqī	占用期
Zhànyòngquán	占用权
Zhànyòng xǔkězhèng	占用许可证
Zhāobiāo tōnggào	招标通告
Zhāobiāo wénjiàn	招标文件

Zhéjiù	折旧
Zhékòu shòufáng	折扣售房
Zhèn	镇
Zhēngdì	征地
Zhēngyòng	征用
Zhěngzhì	整治
Zhèngfǔ jīguān	政府机关
Zhíxíng fǎguī	执行法规
Zhíjiē chéngbāo hétóng	直接承包合同
Zhíjiēfèi	直接费
Zhíxì jiātíng	直系家庭
Zhíxiáshì	直辖市
Zhígōng dàijuànbǐ	职工带眷比
Zhíyè jiégòu	职业结构
Zhìdìng cáiwù jìhuà	制订财务计划
Zhìliàng kòngzhì	质量控制
Zhōngcéng guǎnlǐ	中层管理
Zhōngcéng jiànzhù	中层建筑
Zhōngchǎnjiējíhuà	中产阶级化
Zhōngděng chéngshì	中等城市
Zhōngdī shōurù zhùfáng	中低收入住房
Zhōngguórénmín jiànshè yínháng	中国人民建设银行
Zhōngjiān jìshù	中间技术
Zhōngxīn chéngshì	中心城市
Zhōngxīndìdài lǐlùn	中心地带理论
Zhòngxuǎn biāo	中选标
Zhòngcái	仲裁
Zhōujì gōnglùwǎng	州际公路网
Zhōuzhèngfǔ	州政府
Zhōuzhuǎnfáng	周转房
Zhōuzhuǎnjīn	周转金

Zhòujiān rénkǒu	昼间人口
Zhúbù táotài	逐步淘汰
Zhǔgàndào	主干道
Zhǔyào jiēdào	主要街道
Zhǔyào yòngtú	主要用途
Zhùdì	住地
Zhùfáng	住房
Zhùfáng bǔtiē píngzhèng	住房补贴凭证
Zhùfáng chǎnchūliàng	住房产出量
Zhùfáng diàochá	住房调查
Zhùfáng duǎnquē	住房短缺
Zhùfáng fǎguī	住房法规
Zhùfáng fēnpèi	住房分配
Zhùfáng gǎishàn	住房改善
Zhùfáng gōngyìngliàng	住房供应量
Zhùfáng guǎnlǐ	住房管理
Zhùfáng hézuòshè	住房合作社
Zhùfáng jīdì jìngmiànjī	住房基地净面积
Zhùfáng jiànshè guīhuà	住房建设规划
Zhùfáng jiànzàoliàng	住房建造量
Zhùfáng jīntiē	住房津贴
Zhùfáng jùngōngliàng	住房竣工量
Zhùfáng kāigōngliàng	住房开工量
Zhùfáng liánhé kāifā	住房联合开发
Zhùfáng pǔchá	住房普查
Zhùfáng shāngpǐnhuà	住房商品化
Zhùfáng shēnqǐngquán	住房申请权
Zhùfáng shìchǎng fēnxī	住房市场分析
Zhùfáng xiéhuì	住房协会
Zhùfáng xūqiú	住房需求
Zhùfáng xuǎnzé	住房选择
Zhùfáng yāoqiú	住房要求

Zhùfáng zhànyònglǜ	住房占用率
Zhùfáng zhànyòngquán	住房占用权
Zhùfáng zhuàngkuàng	住房状况
Zhùfáng zhǔguǎn bùmén	住房主管部门
Zhùhù zìyǒu gōngyù	住户自有公寓
Zhùzhái dānbǎo	住宅担保
Zhùzhái zǔtuán	住宅组团
Zhùchǎng gōngchéngshī	驻场工程师
Zhùchǎng jiànzhùshī	驻场建筑师
Zhùlù yòngdì	筑路用地
Zhuānyè shīgōngduì	专业施工队
Zhuānyòng jiànzhù tǐxì	专用建筑体系
Zhuānyòng xǔkězhèng	专用许可证
Zhuǎnhù	转户
Zhǔnbèi shíjiān	准备时间
Zhǔnxǔ dàiyòngpǐn	准许代用品
Zīběn shèbèi	资本设备
Zīběn shuāxīn jìhuà	资本刷新计划
Zīběn tóuzī	资本投资
Zīběn xíngchéng	资本形成
Zīchǎn	资产
Zīchǎn jìngzhí	资产净值
Zījīn quēkǒu	资金缺口
Zīyuán guǎnlǐ	资源管理
Zīyuán huíshōu	资源回收
Zìchóu zījīn	自筹资金
"Zìjǐ dòngshǒu" jiànzhù cáiliào	"自己动手"建筑材料
Zìjiàn gōngzhù	自建公助
Zìjiàn yèzhǔ	自建业主
Zìjiàn zhùfáng	自建住房
Zìliúdì	自留地

Zìrán bǎohùqū	自然保护区
Zìráncūn	自然村
Zìxíngchēdào	自行车道
Zìxuǎn shāngdiàn	自选商店
Zìyíng xiàngmù kāizhī	自营项目开支
Zìzhìqū	自治区
Zìzhù yèzhǔ	自住业主
Zōnghé pínghéng	综合平衡
zōnghéqū	综合区
Zōnghéxìng gōngyè chéngshì	综合性工业城市
Zǒngchéngbāo hétóng	总承包合同
Zǒngchéngbāo rén	总承包人
Zǒnggōngzhǎng	总工长
Zǒngtǐ guīhuà	总体规划
Zūlìn	租赁
Zūlìn gòuzhì xiéyì	租赁购置协议
Zūyòng zhùfáng	租用住房
Zūyuē	租约
Zǔtuánshì jiànshè	组团式建设
Zǔtuánshì qūhuà	组团式区划
Zuìduō jūzhù rénshù	最多居住人数
Zuìjiā yòngtú	最佳用途
Zuìzhōng yànshōu	最终验收

编写《词汇》的点滴回忆

《住房城市规划与建筑管理词汇》(以下简称《词汇》),是中美科技合作协定众多项目中之一个;是根据我国城乡建设环境保护部(即现在的住房和城乡建设部)和美国的住房与城市发展部(下称住房部)的科技合作协定书编写的。这项工作始于1983年,最终完成于1987年。20多年之后,能见到它终于在中国正式出版,作为当时编写人之一的我,在感到无比欣喜之余,还想起了当时工作中点点滴滴的故事。

这本小小的《词汇》诞生在一个有特殊历史意义的年代。在长达数十年的闭关锁国之后,国门开始渐渐地对外开启,1982年美国住房部的塞缪尔·皮尔斯部长应城乡建设环境保护部之邀,率代表团来中国访问,住在钓鱼台国宾馆。我为他们做翻译,也住在那里。记得万里副总理在人民大会堂接见代表团一行的那天，美国驻华大使恒安石也在场，时间正是在中美三个联合公报之一的《上海公报》发表的前夕，所以会见的气氛是极为融洽和愉快的。

万里副总理接见美国代表团一行，1982年
前排左二：美国住房部外事部长助理布立顿；左三：城乡建设环境保护部部长李锡明；左四：美国住房部部长皮尔斯；左五：万里副总理；左六：美国驻华大使恒安石；左七：城乡建设环境保护部副部长张恩树；右一：城乡建设环境保护部科技局局长许溶烈

作为对那次来访的回访，1983年5月，城乡建设环境保护部派出一个由科技局、规划局和城市住宅局三个局局长组成的五人代表团赴美，由美国的住房部接待，那是我第一次踏上大洋彼岸的土地。代表团受到了塞缪尔·皮尔斯部长高规格的接待，在住房部大楼七层他专用的餐厅里设宴款待我们。就是在那次访问期间，5月14日在华盛顿住房部大楼的会议室里由中美双方签署了一份合作文件。

签约双方

前排左坐：美国住房部外事部部长助理布立顿；前排右坐：中国城乡建设环境保护部科技局局长许溶烈；后排左起：钟继光、彭菲菲、约翰·葛瑞蒂、张珑、城市住宅局局长刘辉、规划局局长王凡

内容之一就是共同编写一本包括住房、城市规划和施工管理内容的词汇，目的是为了减少双方交流时因对词义的理解不同而引起误解。

从1983年中美双方签订合作文件起，编写《词汇》的工作就正式启动了。中美双方各组成一个四人工作小组。中方的工作小组由当时的城市住宅局局长林志群任组长，孙骅声、孙汉甫和我有幸被指定参加这项工作。工作的第一步是双方各拟出一定数量的词汇，然后再从中协商筛选。20世纪80年代初，改革开放伊始，对外交流日渐增多，但国内的词典很贫乏，显得很不够用。我们几个人接受这项任务时充满热情，但多少也有点天真，甚至有点傻气。很想借此机会编出一本在建筑业方面真正有用的、内容丰富的词典来。所以居然一下子列出了 3000 多个词条的初稿，而且绝大多数是"软词汇"或"有中国特色"的词汇。所谓"软词汇"是指比较难以理解、难以翻译、并非单纯技术性的词汇。所谓"有中国特色"的词汇，则是因为两国体制和文化的差异，有许多词是对方难以理解，且难以正确地译成英文的。现在回想起来，我们当初傻乎乎地列出的那 3000 多个词条的初稿一定让对方大跌眼镜。因为如果再加上美方选出的部分，其数量和难度将需要一个专业的编写队伍花上多年时间才能完成，是完全不现实的。正因为我们的傻劲，双方花了很长时间经过反复协商和删选，才把入选的词条定在了 1000 个左右，最终定稿是 1036 条。

这项编写工作的特点之一是工作环境和条件如此不同的两个工作组，在远隔太平洋的东西半球分别进行。当时，我们没有电脑，连手动的英文打字机也只是一台老掉牙的。所有的工作全部是手工操作，几近原始。我们为每个词条写一张卡片，一面是中文的解释，另一面是英文的译文，按英文字母排列。当时没有互联网,没有电邮,许多问题或存疑只能以书信方式交换意见,寄一封信到美国大概需要一个星期的时间,等到美方的回音大概一个月也就过去了。完成一部分初稿后则打字出来用快递送达对方。这种"老牛破车"似的交换方式在当今的电脑时代简直是难以想象的。最值得回味的一个例子是我们在华盛顿签署合作协议后仅半年,1983年的10月美方工作组即来京,把编写工作进一步落实。为了那次会谈,我们赶写出一部分词条并翻译成英文。在没有电脑而仅有一台老旧打字机的条件下,要在同一张纸上打出中文和英文相对应的词条是个难题。无奈之下,只能先打出英文,把该写中文的地方留出来,请人用手写的办法把中文写上去。那时候,有一位姓仇的小伙子刚从上海同济大学分配来,是学建筑的,写得一手好字,我们就是请他帮的忙。他在白纸上先用铅笔打好格子,然后用很细的钢笔写上仿宋体中文,其端正程度不亚于印刷的效果。他开了整整一个星期无偿的夜车才完成了这份稿件。发给美方时,他们根本看不出来那中文是手写的。可惜那份原始的初稿没有保留下来,要不然也堪称历史性的文稿了!

《词汇》的另一个特点是词条的解释。在最初的协议里,双方同意仅需对每一个词提供简单定义,不作详细解释。这也是一般编撰词典所遵循的规范格式。但是在我方提供的初稿中,除少数词外,我们都写出了大段详细的解释。因为有些词,例如中国的"基本建设投资"、"房管部门住房"、"户籍"、"四合院"、"借景"等都是对方难以理解,也不是一个简单定义所能说明白的。就说"住房分配"吧,在美国很简单,不是买房,就是租房,没有分配之说。但在那个时代的中国,有单位给职工分配的住房,有城市房管部门分配的住房,还涉及家属宿舍、单身宿舍等,岂能用一个简单的定义来说清楚。同样,美方也有许多词并非简单定义所能说清,如"门槛理论"、"可转让开发权"、"上空使用权"等。我们想既然编写这本《词汇》的目的是为了减少双方交流时因对词义的理解不透而引起误解,就应该说得详细一点。我们的这个大胆创意得到了对方的认同,所以后来全部词条的定义都是按这个口径编写的。绝大多数词条都有大段的解释,仅"建筑法规与标准"一条的解释就长达一页多。这恐怕在词典中是绝无仅有的了。这就是为什么我们最后把它的名称定格为"词汇",而不是"字典";英文名称用的是"Glossary",而不是"Dictionary"。

工作进行到了1984年,我方编写组于9月29日至10月14日赴美工作。目的是双方对大致已经成形的初稿面对面地、逐条逐句地进行讨论，凡是有不清楚的地方可以当面解释，凡是有争议的也可以当面讨论。编写一本中英对照的词典，能和以英语为母语的合作方当面逐一讨论，这种机会也许是绝无仅有的，应该说是这本小《词汇》编写过程中最为亮丽的一个特点。而对于我们从来没有和美国人共同工作过的四个人来说，既是难得的机会，更是巨大的挑战，也是一次难忘的经历。

工作会议在美国住房部办公楼的会议室进行。美方的编写组也是四个人，主编是葛瑞蒂，他是美国住房部的官员。此外，杜凯琳是一位词典学博士，是从美国专利与商标局聘请来的。安德熙是住房部的一位资深专家。李夙炯是美籍华人，曾参加过第二次世界大战，擅长翻译和中西方文化的沟通。此外还有住房部的多位年长资深的顾问，一位特地从美国国防部请来的老资格翻译林先生，还有两位从台湾来的年轻人也一起参加讨论，他们是美方雇来翻译词条的。会议室里还有黑板、投影仪、放映机、录音机等设备。近窗处还放着咖啡、饮料和小点心。总之，济济一堂。我们进去时，着实地感到有点紧张。

讨论分住房、城市规划、施工管理三个主题，逐

中美双方工作组在会议室，1984年

一进行。先提出问题，然后逐条加以解答。美方是经过认真准备的，提的问题都很深入。开始讨论后，双方才发现工作的难度远比预想的大。几乎每一词条都需经过详细的解释方能将其定义翻译得恰当，所以讨论进行得很慢。安德熙说："以前美国住房部与苏联的相应部门也曾编写过一本类似的英俄对照词汇，收入120个建筑技术方面的'硬词汇'，仅给简单的定义，不作详细的解释。而我们这本《词汇》难度大得多，更具挑战性，也更有意思。"讨论越是深入，大家兴趣也越大，在某种意义上几乎成了文化交流了。有些词如果不是当面讨论，其译文肯定会不知所云。记得中方有一个词"施工总平面图"，我们是按字面译成英文的："general construction plan"，但美方却看不明白。结果由孙汉甫到黑板上画了图以后才解释清楚。美方根据她的解释提出了正确的译法"overall construction site plan",竟与我们原来的译法差之千里!谈到"住房分配",那就得从中国的社会主义经济体制说起了。还有中国园林设计方面的一个词"借景",本来美方不明白它的意思,已经把它给否定了。但经过我们解释,却发现它非常有价值,是西方文化中所没有的概念。孙骅声在解释"借景"时,举了借景的经典例子:无锡寄畅园的设计是将远处西山上宝塔的"景"引入了园内,使园内的景色更增添了一个层次。美方经过切磋,认为绝无相应的英文词可用,决定把"借景"按字面直译成"borrowed view",加上引号,并用了大段的解释。在美方提出的词汇中,我们所不理解的就更多了,例如有关房地产和各种形式的抵押贷款等词汇,在20世纪的80年代初对我们都是十分陌生的,翻译也极具挑战性。

给我特别深刻的印象是对"意境"这个词条的处理方法。这个在中国文学、绘画艺术和园林设计中常常用到的词是我们花了许多时间才使对方理解它的含义的。但是美方的几位专家一致认为英文里找不到一个词可以表达"意境"所包含的哲学理念。最后竟然决定用汉语拼音加上引号("Yijing"),然后再加上大段的解释。这个处理方法说明他们对翻译是何等的认真和慎重，是值得我们学习并深刻反思的。如今常常看见许多不负责任的翻译，例如把"地坛"译成"Temple of Hell"，这不仅贻笑大方，而且反映了一种对文化不尊重的轻浮态度。

这样逐条逐句的讨论花了足足两个星期，连星期六也照常工作。星期六是休息日，住房部大楼里空调都关闭，而全楼所有的玻璃窗都是设计成密封式，打不开的。会议室里没有一点通风，又闷又热。但即使这样，大家都工作得非常投入，有争论，有同意，有反对。在认真讨论的时候，往往有人冒出一句幽默的话来，引得哄堂大笑，使严肃的气氛中掺入了轻松愉快的节奏。最后，对所有词条一一举手投票表决，才确定下来最后收入的那1036个词条。我相信任何一本中英双语词典的定义都不可能是经双方面对面讨论并投票表决后才写出来的。若干年后，国内一家有名的出版企业的一位资深编辑见到这本《词汇》后对我说："你们这本词典的编撰方法很独特，很有参考价值。"

中美工作组最后一次共同工作是1985年12月在北京。自1984年回国后，我们在初稿的基础上继续工作，经过了三个月艰苦的努力才完成了我们负责编写的部分,并对美方负责编写的部分进行了校核,发现了问题,准备了该提出的意见。由于有充分准备,所以在他们来访的短短九天中,能顺利地完成讨论,也指出了在他们初稿中的多处错误。安德熙幽默地说:"Who has got an eagle eye?"(你们谁有鹰的眼睛啊?)在讨论中也不乏一些有趣的故事。例如,凡是由美方选出的词条都是由他们负责译成中文的。其中最令人印象深刻的是"brainstorming"。当时我们把它译成"绞脑汁会",现在我常常看见把它直译成"头脑风暴",但在那时候美方的初稿中,竟把它译成了"脑震荡",这就成了一个医学用词了。一经我们指出,美国人都哈哈大笑起来。除了讨论词条外,还讨论了这本《词汇》的"前言"、"编写说明"、"致意"等方面的安排以及最后的出版等事宜。这次会谈后,除了一些扫尾工作外,这一相当具有挑战性的合作项目算是圆满完成了。值得一提的是杜凯琳问我们是否用电脑编辑,我告诉她我们都是用手工操作的,还给他们看了我们所做的卡片和在按英文字母排列时把一条条小纸条贴在大张纸上的原始资料。我们就是用这样的"原始"方法与他们的电脑操作同步进行的,使他们为之咋舌。

编写《词汇》的这段经历已过去30年了,如今回忆起来，当时的许多情景犹历历在目。那一段时间里，我们在美国受到了高规格的礼遇，得到了好评，也交到了朋友，更学到了许多知识。当我们第一次踏进美国住房部的会议室时，感到我们的合作伙伴们是一些既严肃又傲慢的人。但是，共同工作消除了双方之间的隔阂。葛瑞蒂对我们说：这是他第一次接触到中国的知识分子，是非常难得的。以前代表团来了，都是会谈三天，其余的时间都是到各地参观访问。只有我们这个工作组，来美国夜以继日地工作，连周末也没有时间外出游览。我们的敬业精神赢得了同行们的尊敬。他还说这次有那么多人要设家宴款待我们，有的甚至还没有排得上队，也是前所未有的。这里应该插一句：在今日之中国，客人来了，热情接待的方式往往是请他们上馆子狂吃豪饮，但在美国，家宴才是主人对来宾最友好最亲切的招待方式，也是一种最尊敬的表示。所以有那么多人为我们设家宴，让我们感到无上光荣。

家宴有比较隆重的，也有非常随便的。住房部外事部部长助理布立顿（相当于我们的外事局局长）请我们去他家，那天他的太太出差去了，他准备的是简单的冷餐，用的是一次性的杯盘。黄瓜、西红柿等蔬菜都是我在厨房里替他洗净切片的。一顿简单的饭吃了大约四个小时。谈天说地，相互交流，气氛融洽而轻松。告别时，相互的了解增进了，这是最大的收获。隆重的家宴是在麦特卡夫家。这位上了点年纪的先生是住房部资深的住房问题专家。出席家宴的除了我们四个中国人外，还有葛瑞蒂、李夙炯、林先生和苏珊等好几个美国人。那时候才十月初，尚未到感恩节，但他们却以感恩节传统的美食大火鸡来款待我们。偌大的一只火鸡，里面塞满了各种美味，据说必须在烤箱里烤上一整天。切火鸡有专用的刀叉，必须由男主人主刀。除这道主菜外还有苹果饼、蔬菜、冰淇淋等。这顿家宴之隆重还体现在所用的讲究餐具、桌布和所点的蜡烛等方面。饭后大家闲聊，我们向他们介绍中国的唐诗。孙骅声弹月光奏鸣曲，孙汉甫唱歌。他们还夸老孙的琴弹得有举办个人音乐会的水平！另外还有一家的太太在家宴时让我们品尝了一种小红萝卜，说那是她父亲在自家的农场里种的，绝没有化肥的污染。总之，几乎天天有家宴，家家把我们当好朋友来招待。

老资格翻译林先生对我们说：美方在会下议论我们这个工作组有几个特点：一是老林上至政策、下至具体工作都非常熟悉，这样的官员是可敬的；二是四个人都能说英语，熟悉业务，并有很大的工作热情和敬业精神；三是说话坦率，不掩饰自己的缺点或不懂装懂，谦虚、诚恳，有幽默感；四是四个人之间没有上下级之分，也没有男女之分，完全平等地畅所欲言，各抒己见。这些特点和他们曾经合作过的苏联工作组形成鲜明的对照。林先生在美国国防部工作，由于工作的性质，从未回过祖国。但是他视我们为老乡，听到美方对我们的这些好评，就非常深情地告诉我们。

中方四人工作小组，1984年
左起：孙骅声、林志群（组长）、孙汉甫、张珑

这次工作结束后，在合作中所结下的友谊还持续了下去。葛瑞蒂会常常寄些书给我们。在若干年后，我随另一个代表团访美。葛瑞蒂专门为我安排了一次晚餐，把所有当年参加《词汇》编撰的美国同行们都

请来叙旧。记得那天晚餐时,当年美方四人工作小组中最为严肃的安德熙开心地告诉我他的儿子考上了美国西点军校。震惊世界的"9·11"事件发生后,我致信葛瑞蒂表示慰问,并附去当年我们在现在已不复存在的世贸大厦顶层的一张合影。

中美工作组合影,1984年纽约世贸大厦顶层
左起:孙骅声、杜凯琳、孙汉甫、葛瑞蒂、林志群、张珑、李凤炯

他回信说从住房部的大楼里可以看见五角大楼着火。2004年他来信告诉我李凤炯因心脏病去世。这种淡淡的联系虽然无足轻重,但也是我们对那一段工作的一个值得珍惜的回忆。

这本《词汇》诞生在改革开放的初期,我们虽然有幸初次尝试了一次国际合作的经验，但是30年的时间过去了，现在读来，感到里面难免有不足之处，也有不少可以改进的空间。相信随着改革开放的日益深入，国际交流的日益频繁，中美之间的合作项目一定会越来越多，水平也肯定会越来越提高。

张　珑
2013年11月于北京